MAXWELL'S CONUNDRUM

A serious but not ponderous book about

Relativity

Walter Scheider

Cavendish Press
ANN ARBOR

Cover: background photo courtesy of National Optical Astronomy Observatories
Cluster of galaxies in Centaurus, photographed at Cerro Tololo, Chile

Published by Cavendish Press Ann Arbor
Box 2588 Ann Arbor, MI 48106
e-mail at cavendish@worldnet.att.net
further information on this and other books, ideas, and materials for the learning and teaching of the sciences may be found at **www.cavendishscience.org**

Publisher's Cataloging-in-Publication
(*Provided by Quality Books, Inc.*)

Scheider, Walter
Maxwell's conundrum : a serious but not ponderous book about relativity / Walter Scheider. - 1st ed.
p. cm.
Includes bibliographical references and index.
hardcover edition ISBN: 0-9676944-1-8
paperback edition ISBN: 0-9676944-0-X

1. Relativity (Physics) 2. Lorentz transformations.
I. Title. II. Title:Relativity

QC173.55.S34 2000 530.1'1
QBI99-1908

Preface

I am a teacher. Helping people understand difficult ideas is what I do.

The difficulty with Relativity is not the math (high school algebra is just fine); nor are the basic principles complicated or surprising. Relativity is difficult for the novice because it deals with phenomena that are outside the realm of our personal experience. Special Relativity becomes a factor only in events involving extremely high speed, and what happens there is not just a larger version of what we already know. Because it is unfamiliar it seems strange, and often is troublingly counter-intuitive.

To understand Relativity requires above all a trained and disciplined imagination – disciplined because what is to be imagined has to be true! A good teacher, or a good book, can help with that. Most helpful are examples and illustrations that provide food for realistic imagination, consistent and jargon-free vocabulary, careful and complete reasoning, sensible limits on the use of math, a perspective of history, and a sense of humor. I hope you will find them all in this book.

*****

The discovery of relativity changed the way we measure the events of our universe, from the smallest to the largest, in the most profound and fundamental way. This discovery is a story that has captured the imagination of inquisitive minds for nearly a century.

The story that traces this discovery to its beginnings in what seemed like a few innocuous, marginally significant clues is a first order drama of suspense and surprise, of discovery and amazement. Reading about it should be enjoyable and intellectually satisfying.

It is unfortunate that Relativity has accumulated such an aura of impenetrability that readers settle for descriptions of bizarre events that are just never quite explained to them because authors underestimate their ability to reason. The trade has become persuaded that a book on relativity for the non-expert won't sell if it contains equations.[1]

On the other hand, most text books are designed for someone with more time and inclination to plow through dense and copious mathematical arguments than you have.

Most of the so-called "popular" books are not as careful or as serious as you deserve, and most texts are 'way too ponderous. Enjoyment and intellectual satisfaction should not have to come, one at the expense of the other.

That is why this book may interest you. It is written for readers who, whatever their age and occupation, resemble my students who are generalists with intense and wide-ranging intellectual curiosity. They are a diverse group, their professional goals spanning science, medicine, architecture, business, law, philosophy, to name a few. They have in common a sense of wonder about the world around them. Everything they see they want to touch, everything they touch they want to understand, everything they understand they want to try.

Perhaps you have sampled the "popular" books on relativity. Serious in intent, often well written and sometimes quite beautifully done, these books are like birds without wings; they strain to soar, but they don't have the equipment. At critical moments, they typically finesse the hard questions, leaving the logical loops open. They provide unconvincing explanations for what, in the absence of reasons, seem like bizarre and exotic events. They offer a level of enlightenment that Abraham Pais has aptly described as a "nearness to mystery, unaccompanied by understanding."

---

[1] Stephen Hawking reports being advised while he was writing his popular book about time, that "each equation included in the book would halve the sales." Anthony Leggett responded (in a book review in *Physics Today* July 1999 p51) *"No doubt this may be so, but from any but a commercial perspective, a more interesting question is what effect each equation will have on the percentage of readers who will get out of the book the level of understanding intended."*

You may be looking for something that is more serious, without being ponderous.

The inclusion of algebraic logic alone does not make a book serious. "Serious" means thorough and uncompromisingly correct. It tells you where things come from – no facts from out of the blue. Tight, no loose ends, everything with its reason for being.

"Serious" means grown-up, something that doesn't insult the intelligence of a smart person who is not an expert.

Texts, on the other hand, are not merely ponderous, but are increasingly tight and impersonal. Serious contemporary treatises by and large forego the leisure that once allowed scientific authors the literary freedom to converse with their readers.[2] "Not ponderous" means leaving some time and space for such conversation.

"Not ponderous" means not stuffy; not long-winded, making the point, but getting right to it. It means conveying fundamental ideas clearly, while being not ashamed to deal with offbeat and naive questions.

"Not ponderous" means recognizing the puzzling facets of the subject, and dealing with them directly and clearly, using *word-pictures* as well as *picture-pictures*. It means skillfully mixing sophisticated whimsy with serious discourse. It means weaving in the remarkable story of the birth of this idea, without being exhaustively (and exhaustingly) documentary.

"Not ponderous" isn't the same as "no math." There is no way around using some algebra if you are serious about understanding

[2] For some delightful examples of the best in the author-reader conversational style, both past and current, look up Thomas Young's *Interference of Light* Phil Trans Roy Soc London 94 1804; R.B.Lindsay & H Margenau's *Foundations of Physics* John Wiley & Son, London 1936; and Fang Li Zhi and Li Shu Xian's *Creation of the Universe* World Scientific London, 1989. Of current books on relativity, J.A.Wheeler's *A Journey Into Gravity and Spacetime* is superior in this respect.

relativity. But "not ponderous" means not being a mathematical treatise.

A word about mathematics: To be of most use to the broadest readership, this book is designed to move gradually from being largely conversational to being more mathematical. The idea is to enable every reader to reach the highest level of understanding of the subject that his or her mathematical skills allow.

Likewise it moves gradually from the more concrete and tangible toward the more abstract. Rather than beginning with the formalism of spacetime, as is the fashion in current texts, this book begins with observations on the speed of light and moves gradually toward an understanding of how spacetime works and how fundamentally the world of *spacetime* differs from our familiar world of *space and time*.

This approach carries those who can handle a bit of mathematics through a progression that begins by looking at relativistic effects as corrections to classical solutions, and moves gradually to the use of distinctly relativistic energy and momentum equations in the four dimensions of space-time.

This gradual progression gives readers the option of following as far as their math skills permit them to go. However far you go in this book, you will have developed a sound intuitive understanding of the world of relativity, and formed a solid foundation from which to go farther at a time of your own choosing.

The experience offered in this book is within the reach of the non-scientist. With a modest investment of time and effort, any liberal arts student can develop an ***understanding*** of relativity – not merely an acquaintance with it – just as he or she develops insight into the history of the French Revolution and the religious thought of Far Eastern Civilizations.

Contents

Relativity

Richard Schlegel, a physicist at Michigan State University, wrote that, "We study the liberal arts because we are trying to answer certain haunting, fundamental questions... [I would suggest] that physics is the most important of the liberal arts."[3]

The physicists who, in 1948, started the journal *Physics Today* likewise believed that, as the founding editor, David Katcher, put it, "... there is a vast body of educated citizenry walled off from an understanding of physics by its terminology and its disciplines. They are aware of its impact and would like to peer into its depths, be it for curiosity, a feeling that it may have an 'answer' of some sort, or simply because it makes them uncomfortable to have something important go rumbling on outside their ken."[4]

Fifteen years ago I searched and didn't find the book that I thought would be right for my thoughtful and eager students. For them I wrote and duplicated notes that for many years filled in for the book that wasn't there for them.

From those notes evolved "A serious but not ponderous book about Relativity." I hope it succeeds in conveying not only the facts, but also the feel of these phenomena that are so far from the everyday experience of most people. By making these ideas familiar, I hope it dispels the mystique that has grown up around relativity, and succeeds in bringing it, so to speak, down to earth.

This is a journey that my students have traveled with me for many years. This book has evolved through their probing questions and far-ranging discussions, for which I am grateful.

If you care to embark on the same journey by way of this book, I welcome you and wish you well in your travels.

---

[3] Richard Schlegel, the Key Reporter, Spring 1980, p2-4

[4] David A. Katcher, editorial in *Physics Today* **vol 1 Issue 1** May 1948

1.
"Questions that children are taught not to ask"

Einstein's Special Theory of Relativity was born in 1905 out of the realization that three of the very most basic of the known physical principles couldn't all be true. A fundamental incompatibility among the three became apparent. It was a who-done-it with three suspects, each a formidable pillar of unassailable respectability. The conundrum was so perplexing that some of the best detectives in the scientific world were kept chasing clues for more than two decades.

Seen as a tale of mystery, this story has a familiar twist: The detectives have two prime suspects, and the reader sees the spotlight move back and forth between the two, while the guilty party moves among the crowds so inconspicuously, and so totally above reproach and beyond suspicion that even the reader is surprised at how it all turns out. Only this is real life, not fiction.

The First Suspect

The puzzle had surfaced gradually during an otherwise euphoric era in the history of science. For the field of physics, the great theoretical triumph of James Clerk Maxwell in the 1860's had seemed to wrap up electricity and magnetism so neatly that there was talk that perhaps all of nature's secrets had been laid bare and there may soon be nothing left to discover.

An elegant set of four differential equations known as Maxwell's Laws described all known electrical and magnetic effects in terms of Electric and Magnetic Fields. These "fields" were visualized as disturbances in an all-pervasive, invisible medium that was named the æther, in the same way that ripples are disturbances in a water surface and that sound is a disturbance in air.

Maxwell's equations had their dramatic triumph when, in 1887, Heinrich Hertz, using a spark generator, produced a signal consisting of electromagnetic waves while a collaborator captured the energy of these waves some distance away in what could be described as the first radio receiver. This experiment confirmed Maxwell's prediction of twenty years earlier that, like water ripples and sound waves, these electromagnetic disturbances, once created, propagate through space independently of the charged objects that gave rise to them.

It turned out, quite surprisingly, that one can derive from Maxwell's equations not only the prediction of electromagnetic waves, but, even more remarkably, the speed with which these waves propagate through space. This speed is known as "the speed of light," because light is the best-known and most noticeable of electromagnetic waves.

The Second Suspect

This remarkable result from Maxwell's equations was something of a mixed blessing, because it appeared to be incompatible with what is known now as the classical principle of relativity, or the principle of "Galilean Relativity."

The principle of Galilean Relativity says that the laws of motion work the same way within any frame of reference, regardless of any (constant) velocity that this frame of reference may possess. It is this principle that allows us, the inhabitants of a planet that is constantly zooming through space at a speed of 67,000 miles per hour (in its orbit around the sun), to drop things, pour things, throw and catch things, and otherwise live by reasonable and simple laws of motion as if the floor we are standing on were solidly at rest. This classical principle is a cornerstone of Newtonian mechanics, one of the most successful of scientific systems.

If the laws of motion are the same inside an airplane flying at a velocity of 400 mi/hr and one parked on the ground, then it is impossible to say which airplane is moving and which is standing still (except by reference to some marker that defines the background, such as the airport terminal).

Now, what is implied here is that nature can never give us a rule that tells how fast a thing moves, period, because such a rule would then have to apply in all frames of reference, that is, with respect to any marker, whether stationary or moving. Suppose that you are walking at 3 mi/hr along the length of a moving river boat. With respect to a marker on the boat, that is your velocity - 3 mi/hr. The boat, let us say, is cruising down the river at 5 mi/hr. With respect to a marker on a dock at the river bank, you are then moving at a speed of 8 mi/hr. (This is according to a simple rule called the "velocity addition theorem.")

How fast are you moving? 3 mi/hr, or 8 mi/hr? Galilean Relativity says that if you close your eyes, you can't tell. Forget clues, such as the rumble of the boat engine or the rocking due to waves. The laws of mechanics (laws of motion) are no different on the moving boat and on the boat if it were docked. Galilean Relativity, then, implies that no law of motion can tell how fast a thing moves; any such speed is always relative to some marker (called a "frame of reference") that can be chosen arbitrarily.

No one seriously questioned Galilean Relativity. Several hundred years of experimental evidence was in its favor. The chief suspicion was that there may be a technicality by which Galilean Relativity provides a loophole - an alibi - for Maxwell's Laws if light were to be declared not an "object" in the sense in which the Laws of Motion apply to "objects."

The possible loophole aside, Maxwell's Laws appear to do just what Galilean Relativity prohibits; they tell us how fast light travels, period - without reference to a marker.

The reasonable assumption is that the Maxwell speed of light is not universal, and there is an implied, ***fixed marker***, somewhere, that defines by some natural mechanism the unique reference frame in which Maxwell's equations are valid. But there are troubles with that, as you will see.

In any case, being by about two hundred years the newcomer, if not for other reasons, Maxwell's Laws became the prime suspect. The ambiguity over the precise meaning of one of its critical conclusions (the speed of light), helped to point the finger of suspicion in Maxwell's direction. It was Maxwell's conundrum.

Some of the greatest minds of the time struggled with these questions: physicists of the caliber of Holland's H.A. Lorentz and mathematicians like France's J.H. Poincaré.

Experiments, the normal referees of scientific questions were hard to come by because of the extraordinarily great speeds, up around the speed of light, at which the disputed effects occurred. The one exception was the possibility that an experiment could be designed that would clarify the ambiguity over the possible "marker" in the universe that provides the reference frame for the Maxwell value of the speed of light. The experiment was technically extremely difficult, but an American, Albert Michelson, succeeded in doing it.

Michelson's experiment had a surprising outcome. It turned out quite differently from the way it was expected to. Naturally, it was at first thought that this was due to technical flaws in this very tricky experiment. So it was done again and again, each time with greater care and ingenuity. The result was always the same. It failed to find that the earth possessed any velocity at all relative to the fixed marker in the universe that defined the Maxwell reference. It was a great disappointment.

The result was clear, even if unexpected, but few, including Michelson himself, really believed it for a very long time. In retrospect, the result of his experiment was a failure only in the sense in which Columbus' encounter with the Caribbean islands was a failure. Just as it eventually dawned on the explorers that what they had found was more important than a new route to India, after many years the Michelson experimenters came to a similar realization about their finding.

And where was Einstein during this time? Unable to get a job in a university because of his mediocre record as a student, his day job was as a clerk at the patent office in Bern, Switzerland. But his thoughts were engaged in the quest for the answer to Maxwell's conundrum. In characteristic fashion, he asked the questions that others didn't, questions that were dismissed as naive and childish. "Continuing to ask questions that children eventually are taught not to ask, combined with stubborn persistence," is the answer Einstein gave much later when asked to what he ascribed his success in science. As they often are when children ask them, these childish questions were incisive; they cut to the core.

A detective who could find no fault in either of the two suspects

Einstein was well versed in both the mathematical and the physical details that were involved in the Maxwell conundrum. He was convinced that the problem ran so deep, that the solution will be writ large, and he found it difficult to see any flaw in either Galilean Relativity or Maxwell's description of the electromagnetic field significant enough to resolve the disparity.

He saw Galilean relativity as intrinsically compelling, and had little patience with thoughts of carving out a loophole for Maxwell's law based on a technicality.

And Maxwell's equations seemed to Einstein one of those rare giant strides in scientific conciseness and symmetric elegance, describing, as it did, all known electrical and magnetic phenomena, subsuming laws from a century of discovery, laws named after Biot, Savart, Coulomb, Gauss, Ampere, Henry, Faraday, not to speak of Hertz's verification of electromagnetic wave motion.

Like a child playing at the beach with two of the ocean's shiniest, rarest, most precious seashells, being asked which one he would give up, Einstein replied, "Why can't I keep both?"

It was a not a rhetorical question, not an argument. It was a real question. "Just what is wrong with Maxwell's Law just as it is?" Einstein asked. That law says that the speed of light has a certain value. Of course, to be consistent with Galilean Relativity, the speed of light would then have to have that same value in every frame of reference.

Because of the classical velocity addition theorem, this is impossible. The speed of light would have different values measured against markers in different states of motion.

Nonetheless, suddenly Einstein saw the conundrum as no longer a question of whether Maxwell's Laws or Galilean Relativity was flawed. There was a third suspect, that "markers on the boat and on the river bank" theorem, the classical velocity addition theorem, a principle whose venerable record went back not hundreds but thousands of years. Its unchallenged status was based not on a large body of evidence but on ancient common wisdom, the sort that no one but a naive and innocent child[1] would venture to question. Or Einstein.

All of Newtonian Mechanics rested on this common wisdom.

Isaac Newton's early warning

It is to Isaac Newton's everlasting credit that, 200 years before modern relativity, he recognized where, if anywhere, his Laws of Motion might prove flawed. With uncanny accuracy, Newton stated at the outset of the *Principia*, his great work on dynamics,

[1] from Hans Christian Anderson's tale of the Emperor's New Clothes to Frances Burnett's book about the Secret Garden, children have been cast in the role of the unravelers of authority and orthodoxy in thinking.

that there are two unproven assumptions, thought to be obvious, on which the work was vulnerable.

Newton wrote, "The following two statements are assumed to be evident and true:

"1. Absolute, true, and mathematical *time*, of itself, and from its own nature, flows equably without relation to anything external.

"2. Absolute *space*, in its own nature, without relation to anything external, remains always similar and immovable."

Newton did not himself express doubts about these assumptions. But he had dropped a hint, given a warning for posterity that there was an unproven foundation underlying his work, and should for any reason this foundation be found not valid, then all that was built upon it might come tumbling down.

The Unthinkable Third Suspect

Albert Einstein had tracked Maxwell's conundrum to that simple and totally obvious relation that no one ever questioned, the "classical velocity addition theorem." No one had ever questioned it because it was as clearly true as were the two assumptions that Newton had warned us about. In fact the classical velocity addition theorem was true *because of the assumptions that Newton had warned us of.*

That theorem was as sure a thing as the separate and independent existence of the "when" and the "where" of the world, which was to all of existing physics as breathing is to life. It had always been there, essential, unquestioned. To do without it was unthinkable. Einstein was led to think the unthinkable by his reluctance to part with either of those precious seashells. His ability to think unconventionally was one of the characteristics that set Einstein apart.

Einstein's Special Theory of Relativity is nothing more (but also nothing less!) than a declaration of his choice to retain Maxwell's laws, and also to hold on to Galilean relativity, closing any possible loophole in it. So firmly did Einstein believe this to be right, that he was prepared to accept whatever consequences this choice might imply. If Einstein's choice implied abandoning the classical velocity addition theorem, it would mean rethinking the nature of time and space. Which he did.

2. "Because it is so natural and simple"

The Special Theory of Relativity,[1] consists of the assertion of the choice Einstein had made concerning which two of the three principles discussed in the previous chapter were to be retained.

One has to ask what led Einstein to this way of dealing with the apparent contradiction between Maxwell's equations and Galilean Relativity. We will have to look more closely at the essential features of these two principles that made them seem to be incompatible. In doing so, we will be also trying to understand why Einstein found them both so compelling that he chose to question the nature of time itself to preserve them.

Relativity is now universally accepted because of the large volume of experimental verification of its many predictions. But, after all, this mass of experimental evidence was not available to him then, and so Einstein must have been driven by other arguments. The experimental facts that he had to work with consisted largely of those that delineated the apparent incompatibility of the Maxwell equations with Galilean Relativity.

His biographer, Abraham Pais, comments that, "Einstein was driven to the special theory of relativity mostly by aesthetic arguments, that is, arguments of simplicity." [2]

---

[1] The term "Special" describes its domain of application. The "General" Theory of Relativity, which is a theory of gravitation, was published about ten years later.

[2] Abraham Pais *'Subtle is the Lord...' – The Science and the Life of Albert Einstein* Oxford Univ Press, Oxford 1982 p140

It is astounding, at first glance, to think that abandoning absolute time and fixed space would be part of an argument of simplicity. It seems almost impossible to imagine a world in which there is no "Now" that describes the instantaneous configuration of the whole universe at this instant, to be followed by a "Then" which is some time later, everywhere, and so on, past and future.

Einstein was undoubtedly aided in this consideration by the work of mathematicians who had already found that in translating measurements from one station of observation to another, not only distances, but the duration of time intervals involved in travelling these distances might need to be altered. The mathematics was already there in the work of Poincaré, FitzGerald, culminating in the Lorentz Transformation. Unwilling to accept that time intervals might in reality differ in different frames of reference, these workers regarded their mathematics as describing "apparent" effects, which meant that they had reconciled the conflicting principles in appearance only. Yet, the door was opened, as it often is, by a baffling mathematical result whose physical meaning escapes even the creators of the mathematics.

Simplicity

"The history of physics is a history of the reduction of the complicated to the simpler." So said Percy Bridgman, one of the foremost physicists and most thoughtful philosophers of science of the twentieth century.

Science is, by its very nature, a search for common principles from which flow the explanation for multiple phenomena. Both the moon's orbit and the apple's fall are found to be consequences of the law of gravitation. Magnets and lightning turn out to be related to each other and to the aurora borealis and to the deadly radiation of plutonium.

Thus, as we discover more about the world around us, the rules become fewer and more general. Perhaps surprisingly, they do not become more complex as they encompass more diverse events; on the contrary, the most general principles tend to be very simple.

That nature is, at its roots, simple and orderly is not an article of faith among scientists. It is a conclusion drawn from centuries of experience. So consistently has it been observed that progress in science is associated with unification and simplification in our understanding of the basic laws, that this observation itself has become something of a scientific theory, leading scientists to *search* for underlying principles, *expecting* that the ever more fundamental principles will ultimately simplify our understanding of the world.

Simplicity in the basic structure of scientific principles has become not only an expectation, but a test of reasonableness or validity. It is not at all surprising that, for example, in a recent scientific article a particular quantum hypothesis would be critiqued in the following way, "It is ... not at all clear that the theory thus achieved will possess the simplicity and clarity expected of a fundamental physical theory."[3] Scientists feel uneasy about any law that seems needlessly complex.

Simplicity means that a little bit of law goes a long way. It means that a few basic principles, in themselves not complicated, take care of most, or all, the regulation needed to make the world what it is. It means that the most basic laws are simple and sweeping in their generality. Nature is observed to govern not with random and capricious exercise of power, as well she could. For some reason, she is not only dependable and consistent, but also parsimonious. She rules with a few laws of immense scope. Her basic laws are simple in design, even if their consequences may be complex and varied.

And so we should not be surprised to find Einstein charting his way through Maxwell's conundrum using the criteria of simplicity and elegance, sorting and sifting possibilities in his mind with an

---

[3] Sheldon Goldstein, "Quantum Theory Without Observers" *Physics Today* **51** pp42-46 (1998)

incisiveness that was perhaps more the key to his genius than his mathematical skill, which after all was shared by many. "I soon learned to scent out that which was able to lead to fundamentals and to turn aside from everything else, from the multitude of things that clutter up the mind and divert it from the essential," he wrote later.[4]

Galilean Relativity

Galilean relativity carries and honors the name of the Italian who discovered the phases of Jupiter's moons using a telescope he had bought from Dutch sailors. Galileo carved out of medieval ignorance many insights that survived to fuel new physics right down to Einstein. In the hands of Newton they gave us a system of mechanics that by the year 1900 had stood the test of over 200 years of describing how things move.

Although nowadays more powerfully stated in the simple and sweeping principles of conservation of energy and momentum, Newtonian mechanics originally relied upon the concept of force as the agent of velocity change in objects. Forces, Newton's Laws tell us, cause acceleration in objects. Acceleration is the *rate of change* of velocity. You are probably familiar with Newton's Law of Motion,

$$\Sigma \mathbf{F} = m \mathbf{a} \qquad [2.1]$$

This equation says that the *acceleration* (**a**) of an object is proportional to the *total*, or *Sum* of the *Forces* ($\Sigma \mathbf{F}$) applied to it, with the proportionality being related to a sluggishness factor called *mass* (m).

[4] Banesh Hoffman, *Albert Einstein, Creator and Rebel* Viking, New York, 1972 p8

One of the powerful generalizations that followed inescapably from Newton's Laws is that nature makes rules only about ***changes*** in the velocity of objects, not about their velocity itself. Nature provides us no rule from which to determine a value to assign to an object's velocity.[5] This is the essential content of Galilean relativity.

Galilean relativity underlies Newtonian Mechanics. At the same time, Galilean relativity derives its very life from the fact that Newton's Law of Motion is completely tolerant of added velocity, small or large, as long it is unchanging.

Suppose, for example, that your "laboratory" (in the larger sense of the word, as any place in which observations are carried out), is the sailboat of Fig 2.1, moving eastward at a steady velocity u. Any object in your laboratory (the weight being dropped from the top of the mast, for example) will therefore have that laboratory's velocity, u, as an addition to its velocity, v within the laboratory. That additional constant velocity will, of course, have no effect on the *rate of change* of the object's velocity, known as *acceleration*. The object will do the same thing within that laboratory, whether the sailboat is moving or not.

Two things are to be noticed, both really the same - two sides of the same coin.

The first is that, observed in the laboratory of the sailboat, the fate of a weight that is dropped from a perch atop the mast depends not one whit on whether the sailboat is moving (steadily) as in Fig 2.1 or is not moving, as in Fig 2.2. In either case the weight drops straight down along the mast, and falls into the bucket awaiting it at the bottom.

The second is that when observed from a laboratory not moving with the boat, such as the island, the constant horizontal velocity that is now noticeable in the parabolic trajectory of the weight,

---

[5] The equations of kinematics, it might be thought, can be solved to yield velocity: for example the equation $v = v_0 + at$; where "a" is the acceleration. But, in fact, this equation only tells by how much v differs from v_0, how much change v_0 has undergone due to the acceleration.

leaves the *acceleration* of the weight still exactly the same, and still in accord with Newton's Law of Motion, Eq 2.1.

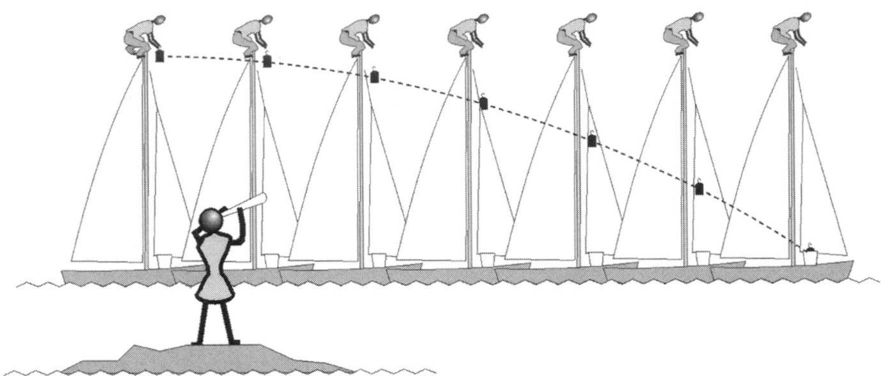

Fig 2.1 A weight dropped from the top of the mast of a moving sailboat falls straight down along the mast to the bucket at the bottom. To an observer on the island, the motion looks like that of a thrown projectile.

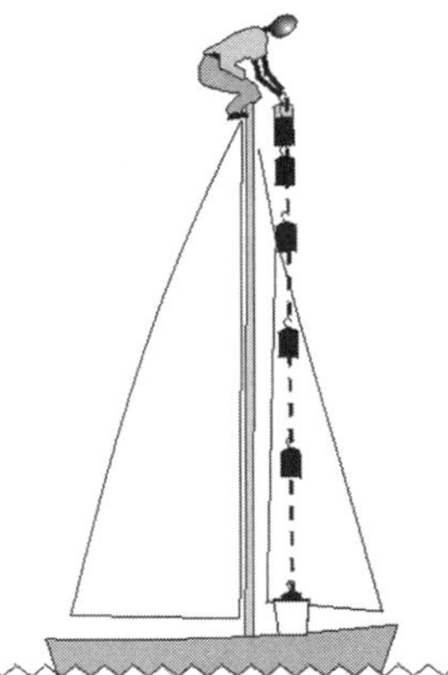

Fig 2.2 (at the right) The falling weight lands in the bucket, whether the boat is moving or not.

The general principle illustrated here is that there is no way for a person on board the boat to determine just by watching the weight drop whether the boat is moving or not. For that, one has to observe fixed things outside one's "laboratory," such as the island. As far as anyone on the boat is concerned, there is no difference between one's "laboratory" being at rest or moving.

You have perhaps sat in a car while waiting for the traffic light to turn green. As you sit, you observe a car in the next lane slowly advancing. Suddenly you realize that the other car is not advancing, but you are slowly drifting backward. You quickly hit the brake to keep from bumping the car behind you.

A laboratory (or frame of reference) is not either moving or at rest except by reference to another object (the island, or the other car). This other object, however, is likewise not either moving or at rest except by reference to something outside itself. An object can be designated as being at rest by common agreement, for convenience, but this does not carry with it anything more than the agreement to designate it that way; there is no universal or absolute meaning to being at rest. There is only relative motion, one object with respect to another, one laboratory with respect to another. There is no special laboratory that is in some absolute sense ***the*** reference laboratory for the universe, none that is "fixed" while others are "moving." This is the principle of "classical relativity," or "Galilean relativity."

A "laboratory" in the sense we have used the word here is called a "frame of reference." A ***frame of reference*** may be thought to be the x,y,z axes along which distances are measured.

It is easy to see that this principle is of great practical importance to us. We live our lives in a "laboratory" that is moving at a speed of 67,000 mi/hr, the earth's speed in orbit around the sun. If this movement had any detectable effect on how objects behave on earth, that is, on the laws of motion on earth, it would likely complicate our lives greatly (Fig 2.3).

Especially is this so because every six months we move to the opposite side of the orbit, where the direction of the earth's velocity is just reversed, corresponding to a *change of velocity* of 134,000 mi/hr. But, our laws of motion here on earth seem to be quite unaffected by such a great change in the velocity of our frame of reference, a comforting reassurance that they are consistent with Galilean relativity.

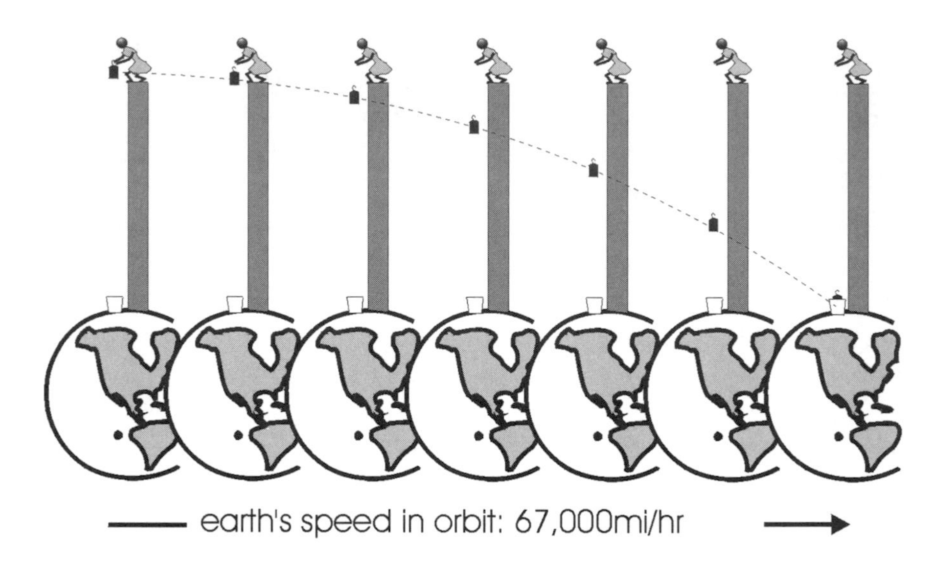

Fig 2.3 (above) Galilean relativity in action: the weight falls "straight down" unaffected by the earth's motion at a speed of 67,000 mi/hr.

Fig 2.4 (right) Motion occurs on earth as if earth were standing still. Only by observing the sun and stars can we determine the nature of the earth's motion. (The curvature of the earth's path is being neglected here; its effect on the motion of objects on earth is small.)

Maxwell's Equations and the speed of light

Rather unexpectedly, in deriving the prediction that Electric and Magnetic Fields can propagate as electromagnetic waves, Maxwell found that the equations told also ***how fast*** these waves would propagate. This speed was not a variable that would be affected by the velocity of the laboratory in which the speed of the waves is being measured; it came out as a number, a very specific number. Its value is about 3×10^8 m/s.

This number was not a totally unfamiliar quantity. It is also the number obtained by various experimental scientists who had by various methods actually measured the speed of light. Dating far back in time, light had been clocked making trips around the universe and between mountain tops, in spite of considerable disagreement about what light is.

The earliest estimates of the speed of light were obtained from astronomical observations as early as 1676 by Ole Roemer. Later, the speed of light had been calculated from an aberration in the apparent position of stars due to the earth's movement sideways to the approach of the light (James Bradley, about 1740). In 1849, H.L. Fizeau reported a measurement of the speed of light obtained by bouncing pulses of light off mirrors on mountain tops. All of these measurements were in general agreement with each other, and had placed the speed of light close to 3×10^8 m/sec.

It was suspected immediately that it was no coincidence that the numbers found by these experimenters were similar to the value of the propagation velocity of electromagnetic waves predicted by Maxwell's Laws, and that this similarity pointed to the identification of light as an electromagnetic wave phenomenon. This particular number, though it applies to all electromagnetic waves, was soon called the "speed of light." Einstein referred to Maxwell's finding as "the law of the propagation of light."

The discovery that Maxwell's radio signals travel at the same speed as light waves suddenly enlarged the scope of Maxwell's Laws beyond the traditional realm of electricity and magnetism, to include also the whole discipline of phenomena associated with light. Because it came unexpectedly out of the equations dealing with electric and magnetic fields, this annexation of what had been thought a totally different area of physics was a much celebrated event.

It was also, however, the event that raised the curtain on the drama of the conflict between Maxwell's equations and Galilean relativity. The discovery that Maxwell's equations predict the speed of light in fact became Maxwell's conundrum.

You will immediately recognize that this sudden interest shown by nature in what the velocity of something ***is***, irrespective of the velocity of the laboratory, is new, and is forbidden under Galilean relativity.

Galilean relativity owes its existence to that aspect of Newtonian Mechanics which causes the Law of Motion to be indifferent to the addition of any constant velocity attributable to the motion of the "laboratory," or frame of reference. The acceleration of a moving object ***is*** independent of the speed of the laboratory in which the instruments of measurement are located, but its velocity ***is not***, which explains why Galilean Relativity permits laws determining an object's acceleration, but not laws determining its velocity.

For example, the velocity of a sleeping passenger in an airplane flying at 400 mi/hr would be described as zero by a fellow passenger, but would be said to be 400 mi/hr by an observer at the airport terminal. The difference in the reported velocities has no impact on Newton's Law of Motion, because it has no impact on the values of acceleration. But no law of nature can say whether the passenger's speed is zero or 400 mi/hr.

What to make then, of Maxwell's equations that give a speed of light without reference to a marker, "fixed" in the universe. Perhaps such a special frame of reference does exist, after all, and is a part of Maxwell's equations by implication.

It was presumed that the value of the speed of light that is derived from Maxwell's equations is the value that one would determine in an actual speed-of-light measurement in such a unique, special "laboratory." But what laboratory? What frame of reference?

Because the Maxwell value fit so well with the results of measurements made by various experimenters, one naturally looked among those experiments for that special frame of reference. Because some of these experiments were earth–bound while some were astronomical, the answer was anything but clear.

It was assumed that a favored frame of reference exists in which Maxwell's equations are "exactly true," in spite of the fact that this assumption is a violation of Galilean relativity. It would follow, then, that in any laboratory that is in motion with respect to that favored frame, the speed of light would be different from that predicted in Maxwell's Laws. Our fast-moving earth would be an example of such a laboratory. How then account for the fact that Maxwell, using the results of experiments done on this earth, found a law that specifies the speed of light in some special frame of reference that is not our own?

The earth's velocity in orbit, 67,000 mi/hr (about 3×10^4 m/s) is 10,000 times smaller than the velocity of light. Perhaps, it was thought, the minor effect of this small movement of the earth laboratory is just not noticeable, and can be ignored.

Astronomers did not know about galaxies at the time, much less the fact that some of these galaxies travel at half the speed of light. So the discussions of that era about Maxwell's conundrum were oblivious to how unusually lucky Maxwell would have had to be to live in a galaxy that did not add a huge galactic velocity to the 67,000 mi/hr orbital speed of his earth-bound laboratory!

In any case, these arguments were unconvincing to many serious workers in this field. It was, after all, not a matter of whether the effect is small or large, but that there is an effect at all. That there should be a special frame of reference in which Maxwell's equations are exactly true, whereas in others they are not, is not permitted by Galilean relativity.

A frame of reference is usually associated with some object, or "stuff" of some sort, to which one can attach (even if only conceptually) x, y, and z axes. The value of the speed of light that is obtained from Maxwell's equations is presumably of universal significance, so this special frame of reference must span the entire universe. Just what is it that defines this special frame of reference; what are those x, y, and z axes attached to that keeps them anchored in the universe, that keeps this huge laboratory fixed? In the next chapter we will examine these questions, as well as the one aspect of them that a determined pair of experimenters succeeded in tackling: how fast is the earth moving through this fixed frame of reference?

The Loophole in Galilean Relativity

No one seriously questioned Galilean Relativity. It was, after all, a statement about the Laws of Mechanics, and everyone was certain what those Laws were. These laws about the motion of objects were, in fact, the same in all laboratories whose motion, no matter how great, was constant.

This left the problem of the violation of Galilean Relativity by the presumed interpretation of Maxwell's Laws. This question attracted the quibblers, the nit pickers. Are Maxwell's equations ***Laws of Mechanics***, or are they not? True, they deal with the motion of things, but isn't it possible that light and other electromagnetic waves are so fundamentally different from moving "objects" that Galilean Relativity does not apply to them? If so, then Maxwell's law of the propagation of light may just fall through a loophole in classical relativity.

Einstein didn't buy this argument for a moment.

Here is his opinion, "Even though classical mechanics does not supply us with a sufficiently broad basis for the theoretical presentation of all physical phenomena, still we must grant it a considerable measure of 'truth,' since it supplies us with the actual motions of the heavenly bodies with a delicacy of detail little short of wonderful. The principle of (classical) relativity must therefore apply with great accuracy in the domain of *mechanics*. But that a principle of such broad generality should hold with such exactness in one domain of phenomena, and yet should be invalid for another, is *a priori* not very probable." [6]

With these words Einstein spoke not only for retaining classical relativity, but for explicitly plugging any and all loopholes that might provide for exceptions. This sharpened the conflict between relativity and Maxwell's law of the propagation of light, for, to Einstein, relativity now explicitly excluded the possibility that *any law of physics* could specify the value of the speed of light in one particular frame of reference alone.

Just in case you are still prompted to wonder, Doesn't the weight in Figs 2.1 and 2.2 fall straight down in only one frame of reference, the one in Fig 2.2 in which the sailboat is at rest? Isn't that frame of reference then "special?" The answer is, "No." Galilean relativity is not about trajectories; it is about laws of motion. The Law of Motion is the same in both frames of reference; the Law of Motion predicts straight down motion in one case and parabolic descent in the other. The indifference of Newton's Law of Motion to added constant velocity accounts for that.

Maxwell's equations, specifying a particular value for the *velocity* of light, do not have this generous indifference to laboratory motion.

Here, then, is the conflict. Einstein, at this point, accepted that in closing the loophole in Galilean relativity, he had exiled Maxwell's equations from the kingdom of classical relativity. It appears, Einstein commented, that one must either abandon the "simple law of the propagation of light *in vacuo*," or one must abandon the (classical) principle of relativity. "Those of you who have carefully followed the [foregoing] discussion are almost sure to expect that we should retain the principle of relativity, which appeals so convincingly to the intellect because it is so natural and simple," he added.

---

[6] A Einstein, *Relativity* Crown Publishers, NY Orig publ 1916, reprinted 1961 pp13-14

Einstein had painted himself into a corner. He had initially agreed that he could not keep both of those shiny seashells. He had committed himself forcefully to the widened principle of relativity, which would have forced him to yield on Maxwell's Laws. Yet he felt almost as strongly about what he called the law of propagation of light. We end this chapter not knowing how he will extricate himself from this conundrum.

3. Maxwell's Conundrum

The conundrum is not yet solved. We have to understand it better. It is subtle.

Maxwell's conundrum arises out of the fact that Maxwell's equations lead to the conclusion that electromagnetic signals propagate like waves and do so at a speed of 3×10^8 m/s. The presence of this number in a law of physics is what makes Maxwell's equations fundamentally different from Newton's Law of Motion, as regards consistency with Galilean Relativity.

All this harping about whether or not nature can specify the speed of something can seem like a lot of fuss about what may appear to be a fairly commonplace event. Isn't nature constantly in the business of determining how fast something travels?

A truck is moving at 75 mi/hr. It will *seem* as though the truck is moving at that speed if you are in a stationary patrol car; it will *seem* as though it is moving at 10 mi/hr if you are in a patrol car that is chasing it at 65 mi/hr. But the truck is moving at 75 mi/hr.

Sound travels (in air at ordinary temperature and pressure) at a speed of 750 mi/hr. If you are moving toward an oncoming sound at 60 mi/hr, the sound will *seem* like it is approaching you at 810 mi/hr; this has a noticeable effect on the pitch of sound waves, a phenomenon known as the Doppler effect. But sound travels at a speed of 750 mi/hr.

How is the conclusion from Maxwell's equations that light is propagated at 3×10^8 m/s any different? Could we not all agree that this is the speed of light, just as we can all agree that the speed of the truck is 75 mi/hr and the speed of sound is 750 mi/hr?

This is, in fact how most scientists viewed the meaning of Maxwell's law of the propagation of light at first. The cause for some unease about this view arises out of the fact that, on a rapidly moving planet or in a rapidly moving galaxy, not only would the speed of light *seem* to be different due to the motion, but it would *seem* to be not the same in the forward and the backward direction.

Furthermore, because the value of the speed of light comes directly out of the Maxwell equations, these equations themselves would have to *seem* to be different in order to predict the *seemingly* different speeds of light. Now, what kind of physics would that be, in which the laws of physics take on a different form East-to-West and West-to-East?

Even if all this were so, there would be the need to explain why, to us here on Earth moving at a speed of at least 67,000 mi/hr (augmented possibly by even greater velocities attributable to the solar system as a whole and to our galaxy), the laws of electromagnetic fields do not do not appear to be modified as a result of such a frame-of-reference adjustment. Right here on earth these laws seem to work perfectly well just as they are.

A coincidence? Does it mean that we are favored with being at rest in that unique special frame of reference in which Maxwell's equations work as they are written? An unlikely happenstance in any case, but almost impossible in view of the difference in our velocity through space at different points in our solar orbit.

The question is: Is the speed of light in Maxwell's law a number that applies exactly only in some fixed frame of reference, just as the speed that we agree to call the "actual" speed of the truck is what is observed from a marker at the side of the highway?

The enormous speed of light makes the question about light a good deal more difficult to answer than the corresponding question about the truck.

How Maxwell's equations give us the speed of light

We have noted that the speed of propagation of electromagnetic waves emerges from Maxwell's equations without any actual measurement of that speed.[1] The speed of propagation of these waves is determined completely by two quantities that can be found from static measurements, where there is no wave and nothing at all moves. Where did these equations come from?

Maxwell had obtained his equations by restructuring the mathematics of electromagnetic laws of Faraday, Ampere, and Gauss,[2] which are in integral form, into a form in which they can be solved giving new equations that turn out to be equations of wave motion. These predict that once somebody has wiggled an Electric or Magnetic Field (by wiggling some electrons, as, for example, in a transmitting antenna), the field develops a life of its own and continues to wiggle off into the universe like a long rope that has been snapped at one end.

Maxwell's equations showed that changes in electric or magnetic fields will propagate through empty space. Vibrating electromagnetic fields propagate as waves, just as sound and water waves do.

20 years before Hertz made his radio waves in a laboratory, the solution of Maxwell's equations had predicted not only the existence of such waves but also the speed of their propagation.

The only information needed in order to calculate the speed of propagation of these waves was already available from static or steady-state measurements of: 1) the dielectric constant of space, ϵ_o, using a charged capacitor and 2) the magnetic permeability of space, μ_o, using a wire carrying a constant current.

---

[1] A derivation of the speed of propagation of electromagnetic waves from Maxwell's equations appears in Appendix I.

[2] Maxwell's equations include certain terms that are sometimes omitted in elementary treatments, but which are necessary to make the equations generally applicable.

In Appendix I, this speed, which is given the universal symbol, 'c', is shown to be related to the dielectric and magnetic constants in the following way:

$$c = 1/\sqrt{\epsilon_o \mu_o} = [8.85\times10^{-12}\,\mathrm{F/m} \times 4\pi\times10^{-7}\,\mathrm{N/A^2}]^{-1/2} \quad [3.1]$$
$$= 2.998\times10^{8}\,\mathrm{m/sec}$$

When he noted the similarity of this number to measured values of the speed of light, it naturally struck Maxwell that light waves could actually be ***electromagnetic waves***. (Michael Faraday had speculated about this possibility earlier, but with no substantial evidence.)

That light is wave motion of some sort was the gist of a feisty report given on November 24, 1803, sixty years before Maxwell's breakthrough, by Thomas Young standing before the Royal Society to tell about an astoundingly simple experiment that he had done with sunlight, a mirror, and a couple of little "slips" of cardboard. That experiment had substantially removed any doubt that light was a wave phenomenon.[3]

And now that static measurements enabled scientists to predict how fast these waves propagate, there was an increased urgency to know more about just what a light wave is and how it propagates.

The supposition was that being a wave means that it is a disturbance in something, just as a sound wave is a disturbance in air. But what is it that a light wave is a disturbance ***in***? Presumably whatever that stuff is will turn out to be what holds that special frame of reference in place in which 2.998×10^{8} m/s is the speed of light.

---

[3] Thomas Young, *Phil. Trans. Royal Soc. London* **94** (1804); reproduced in H.Crew, ed., *The Wave Theory of Light* American Book Co, New York (1900) and in M.H. Shamos, ed, *Great Experiments in Physics* Holt Rinehart & Winston, New York (1959).

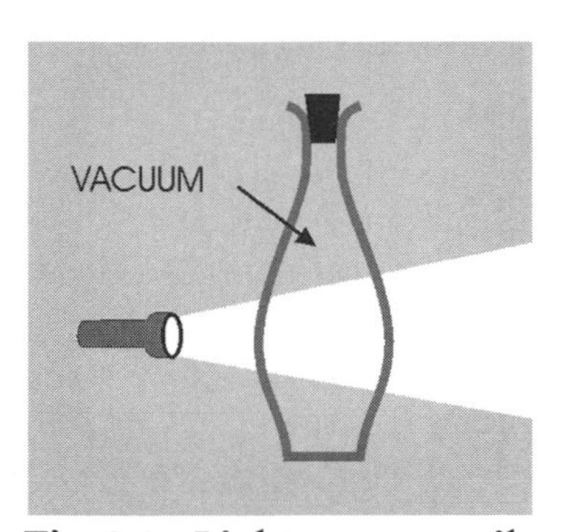

Fig 3.1 Light passes easily through vacuum. Pumps seem unable to pull æther out of the bottle.

The stuff that light rides on

What the material might be in which electromagnetic waves travel was difficult to understand. Light (an electromagnetic wave) passes with ease through the vacuum in a glass bottle from which all gases have been pumped, so, whatever that material might be, it is present in what we normally identify as vacuum. Light also passes readily through the open space between planets and stars, where atoms are very scarce. Yet the leading scientists hung tenaciously to the belief that if there are waves, there must be *something* upon which the wave disturbance exists and which accounts for their propagation.

For want of knowing what it is, it was described as an invisible, sparse, gelatin-like, elastic material, apparently able to pass easily through solids (vacuum pumps seemed unable to pull it out of evacuated bottles, so either the glass of the bottles or the steel in the pumps is porous to the material on which the light waves ride). This will-o-the-wisp substance was described as "etherial" and was called "æther," with the "æ" a bit of pretentiousness that also helped to distinguish it from the organic solvent called "ether."

The æther hypothesis gave support to the idea of a material that fills the universe and is anchored in some way to the great beyond, defining what "at rest" means. In this gelatin-like background material, surely one could draw axes as in a three-dimensional graph, against which it could be determined how fast something "actually" moved.

This gave to the value of '*c*' from Maxwell's equations a simple meaning. '*c*' is the velocity of light in the æther. The æther was the "medium" in which all electromagnetic waves are propagated.

The Velocity Addition Theorem – the unlikely culprit is unmasked

Einstein had asked this question: just what rule is it that does not permit the principle of relativity to live side-by-side with Maxwell's law of the propagation of light? Here, then is the answer: In a laboratory *not* fixed in the frame of reference of the æther, the speed of the laboratory would have to be added or subtracted (as a vector) to 3×10^8 m/s to obtain what would *seem* to be the speed of light in that laboratory.

If the speed of light *seems to be* not exactly 2.998×10^8 m/s, in a laboratory moving with respect to that special frame of reference in which the æther is at rest, then Maxwell's equations violate Galilean relativity, because they constitute a law of physics that is not the same in the two frames of reference.

The speed of light would, by some unknown mechanism, have to be the same in all frames of reference for Maxwell's equations to live side-by-side with relativity. But this was not considered possible, any more than the speed of the truck could be the same from the point of view of both moving and stationary patrol cars.

The principle that keeps this from being possible is that "markers on the boat and on the river bank" theorem, more properly called the "classical velocity addition theorem." It is usually explained to students in terms of a common sense example like that of the truck, or of persons riding an escalator upon whose steps they are climbing even while the escalator is moving upward.

This common-sense logic begins with the obvious fact that every velocity is measured relative to some scale of distance and location that we have called a reference frame.

For example, when you are seated in a moving car, a fellow passenger would say that you are at rest; your velocity relative to the car is zero. An observer whose frame of reference is fixed by markers at the side of the road would claim that your velocity is not zero, but is the same as the velocity of the car.

Imagine that a person is climbing up a moving escalator. Although the person could reach the next floor by walking up a flight of stairs that is not moving, the moving stair helps that person reach the top faster. The resulting "actual velocity," which could be measured by an observer standing on the floor below, is simply the sum of the *climbing velocity* and the *velocity of the escalator.*

Fig 3.2 The velocity of the person (with hat) depends on who is making the measurement. Illustrates the classical velocity addition theorem.

In Fig 3.2 the escalator stairway is the moving frame of reference; let us refer to its velocity as u. The climbing person is clocked by the little observer standing behind him on the escalator, who finds the climber to have a velocity, v, relative to the escalator.

The observer on the floor below the escalator is said to be in a fixed frame. That observer finds the climbing person's velocity to be V. "Common sense" tells us that the relation among these three velocities is,

$$V = u + v \qquad [3.2]$$

Suppose that the escalator is moving with a speed, $u = 2\,\text{m/sec}$, and that the climbing rider is walking up the escalator steps at a speed, $v = 2\,\text{m/sec}$. We would then judge that the rider is "actually" approaching the top of the staircase at an "actual" velocity, $V = 4\,\text{m/sec}$.

The concept of an "actual" velocity assumes that there is some frame of reference that is by general agreement designated as fixed, such as, the floor below the escalator. (Of course, the floor, being part of the earth, which rotates and orbits the sun, is in the larger sense not fixed.) For the purpose of observing moving escalators, the earth is sufficiently fixed. We will see later why this is so.

Eq 3.2 is called the ***classical theorem of the addition of velocities***. It is often mistakenly regarded as a mathematical identity. It is actually a fact derived from certain assumptions about space and time, since, after all, it deals with velocities, and velocities are relations between a distance moved and the time required for that motion to occur.

This theorem is regarded as "self-evident," because it is usually taken for granted that distances in moving and fixed frames can be measured by the same scale - that, for example, 2 meterstick lengths along the floor is unconditionally equal to 2 meterstick lengths along the moving stairway, so that these two distances can be added as if they were laid end to end in one and the same reference system, giving exactly 4 meters. This theorem was so well accepted because the condition just described seems so reasonable.

Being able to add these two distances is not enough; to find velocities one must also be able to divide each of them by an equivalent amount of time. It is therefore also necessary that one second measured on a clock in the fixed frame is unconditionally equivalent to one second measured on the moving escalator.

To be able to add u and v, the velocities on the right side of Eq 3.2, these assumptions about distance and time have to be true.

It is, as you see, not entirely simple. An amazing experiment intervened in the consideration of this matter.

An experiment to find out how fast the earth is moving through the æther

If the earth were in motion with respect to the æther, the measured speed of light would be different in different directions, just as the speed of sound in moving air is different with and against the wind, and the speed of a swimmer in a flowing river is different upstream and downstream.

It seemed clear, then, that the earth's motion through the æther could be determined by measuring how the speed of light varies with direction. There was considerable theoretical interest in determining how fast the earth is moving through space. Since the earth's movement through the æther would be perceived by an earth observer as a movement of the æther past the earth, the speed whose value was sought was called the "æther drift."

Among those with curiosity about this question was, naturally, Maxwell himself. The experiment would be difficult if the speed of the earth through the æther were small compared with the speed of light. Discouraged about the feasibility of carrying out a totally earthbound light experiment with sufficient accuracy to reveal the earth's motion, he suggested research on light from Jupiter's moons, which in their orbits move alternately toward and away from the earth. This, he believed, would reveal the effect of the motion of a source of light on its speed. Maxwell reviewed the situation in a letter in 1879, shortly before his death. In 1881 a U.S. naval officer on leave working in Helmholtz's laboratory in Berlin, having read Maxwell's letter, remained convinced that he could devise an earth-bound apparatus that could measure the effect of the earth's motion on the measurement of the speed of light. His name was Albert A. Michelson; the apparatus he built is now called the "Michelson interferometer."

Michelson did the experiment and his result disappointed him. He did not find what he was after. The experiment, designed to determine the speed of the earth as it zoomed through the universe, came up blank;[4] the speed of the earth was zip, zero. Just in case the experiment happened to have been done at a moment when the earth was temporarily "at rest" in the universe, he repeated the experiment but obtained the same result.

Michelson was dissatisfied, and, a few years later, did it again, with more elaborate care, and with a collaborator, Edward W. Morley. The experiment was in theory feasible, they reasoned, because the earth can not be at all times at rest in the universe's fixed frame. If earth happened in April to be at rest in the universe, then in October it would not be. This is because the earth makes a big circle around the sun once a year, and is moving in opposite directions on the opposite sides of its orbit. The earth's speed in orbit is about 67,000 miles per hour. This speed is only 1/10,000 the speed of light, which made the experiment technically formidable, but not impossible.

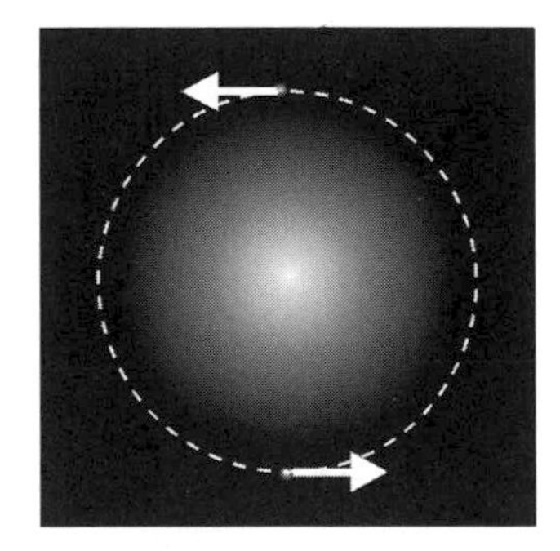

Fig 3.3 The earth's velocities at opposite sides of its solar orbit are reversed.

If, for example, the sun were standing still in a fixed universe,[5] then light would travel, at some time during the year, at 297,970,000 m/s if projected toward the west and 298,030,000 m/s if projected toward the east. If the speed of light could be measured accurately enough to distinguish these speeds, the answer to the question would be attainable.

Fortunately it was possible to measure the difference between the speeds without having to measure the speed of light itself with that kind of accuracy. The scheme of the interferometer was to split a light beam in half and set up a race between the halves. Whether the two halves arrive at the finish line

[4] Abraham Pais *'Subtle is the Lord...' - The Science and the Life of Albert Einstein* Oxford Univ Press, Oxford 1982 Chapter 6.

[5] This was at the time still a plausible supposition.

together, or whether one arrives ahead of the other (even *how far ahead* of the other) can be determined to great precision by observing *interference* patterns produced by the overlap of the highs and lows of the wave patterns of the two half-beams.

The beam of light is split by sending it through a thinly silvered mirror, which passes half the beam and reflects the other half. The mirror can be placed so that the reflected portion is sent off at 90 degrees from the incoming path. Both portions of the split beam are then reflected back toward a common point after traversing some distance along their respective paths. When the reflected beams recombine, they *interfere* with each other. "Interference" is a term used to describe the cancellation that occurs when wave "tops" come together with wave "bottoms," and the addition of wave heights that occurs when they come together with other wave "tops."

If the speed of the light in one of the half-beams were to be even very slightly different from that in the other, the interference pattern would undergo a noticeable shift. This is because the wave length of light is so very small (about 5 millionths of a meter), and even a small delay in one beam can cause a lag of a fraction of a wave length in its arrival at the common point.

It is, of course, not possible to make the two paths exactly equally long. But for the purpose of the experiment, that is not necessary. By making the paths at right angles to each other, and rotating the entire apparatus, a position can be found in which the motion of one beam is *along* the direction of motion of the earth and the motion of the other beam is at right angles to the earth's motion.[6]

By placing the interferometer so that first one beam is along the motion of the earth, and then rotating it so that the other beam is

---

[6] It is as if two swimmers in a river each swam from a particular point a distance of a quarter mile and back, one swimming first with and then back against the current of the river, and the other swimming at right angles to the current. It is not obvious, but can be shown that the swimmer who swims first with and then against the current takes longer to make the round trip than the swimmer whose path is perpendicular to the current.

along the motion of the earth, first one beam and then the other would be slightly delayed, and the *shift* in the interference pattern would indicate the speed of the earth in the universe's at-rest frame. By doing the experiment in April and again in October, one can be assured that in at least one of the experiments the earth will not have been at rest with respect to the universe's frame. In short, the device made possible a critical test of the hypothesis that there is in the universe a particular frame of reference that is fixed and at rest. This was presumed to be the frame in which the measured velocity of light would be the one given by Maxwell's equations.

A most remarkable "failure"

The experiment performed in 1887 by Michelson and Morley, still gave a "null" result. Even though many other experimenters repeated the experiment (with mixed results), and some questioned the null result as late as 1933, the scientific community had accepted the puzzling outcome and in recognition awarded Michelson the Nobel Prize in 1907.

At all times, day and night, April and October, the velocity of light measured on earth appeared to be in all directions the same.

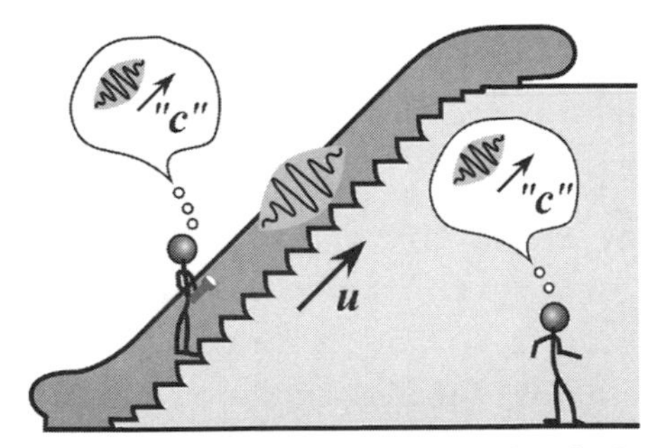

Fig 3.4 Both observers see the *light pulse* travel at speed "*c*." The escalator velocity does not add.

By analogy, it is as if a swimmer were seen to move at the same speed with, against, and across the water current in the river. It is as if the movement of the escalator that adds speed to the human rider, adds nothing to the propagation speed of a pulse of light beamed up the escalator from a flashlight held by the rider (Fig 3.4).

One should not imagine that the implications of this experiment were immediately apparent. Even Michelson remained firmly attached to the idea of an æther fixed in the universe. At that time, galaxies were unknown, and the stars were thought to be fixed in the sky. In 1902, fifteen years later, Michelson was still apologetic about the failure of his experiment to elucidate the earth's velocity in the universe. The idea that the universe could be without an absolute reference frame, and that this experiment provided the proof, seemed unlikely even to the two experimenters.

It is important to recognize that the ultimate implication was that the universe is without an absolute reference frame not by circumstance, but on principle.

What *seems* and what *is*

It *seems*, then, that regardless of how fast or in what direction you (and all the instrumentation needed for measuring the speed of light) are flitting about the universe, and regardless of the motion of the source of the light, your instruments will give you the same answer for the speed of light.

This is exactly the interpretation that Einstein was willing to give to the Maxwell speed of light, because it is what Galilean relativity required. He found he had to give up the classical velocity addition theorem to do this, and along with it the assumptions of absolute time and fixed space. Others were not willing to sacrifice these assumptions.

Michelson's own apology for this result alludes to the expectation that there will be some way to explain this result away, because, of course, in the world of absolute time this is not supposed to happen.

The ability to explain something away is the crucial difference between "seems" and "is." In a field like relativity, where so much revolves around "observing" and making "measurements," it is important to define the distinction between what seems and what is. *Seems* means that there is an appearance or illusion due to some well understood principle that allows one to extrapolate from what ***seems to be*** to what actually ***is***.

What led Michelson, and many others for thirty-some years after the experiment, to continue to doubt the result of his experiment was the belief that it is an illusion based on some effect that will eventually be revealed. The persistence of their doubts is testimony to the hold that the classical, or common-sense, velocity addition theorem had on the body of common wisdom.

According to that theorem, any motion that you (and your instruments) have with respect to a moving light wave would affect what would ***seem*** to you to be the speed of propagation of that light wave. The Michelson Morley result is not consistent with that theorem.

The story is told that at 16, Albert Einstein was curious about what light waves are really like. He imagined being able to fly through space at the speed of light, and line himself up with a moving light wave. What would that light wave look like if he were abreast of it? [7] Only later did this turn out to be the crucial question that cracked open the door to relativity.

Common wisdom suggested to the young Einstein that the light beam would neither gain nor lose on his position with respect to it, and so would appear to be standing still. He understood that the light wave standing still would be an illusion. Knowing that ***he*** was moving at the speed of light, he could correct the illusion and conclude, as would be expected, that the light beam does indeed travel at the speed of light. Why, then, was this scenario a puzzle to Einstein?

Aside from the fact that it is not clear what a standing-still light wave might look like, did he already then have a suspicion that there is something wrong with the whole picture? He brooded on

---

[7] Moving abreast of a travelling wave is not such an absurd thought. An airplane does this as it breaks the sound barrier, when it is travelling at the speed of sound. The airplane is travelling right alongside the leading edge of the sound waves produced by its own engines. The shock waves and "sonic boom" that are observed are well understood. But sound and light are fundamentally different in this respect.

it, suggesting that perhaps even then he may have harbored the germ of a doubt about whether classical velocity addition would apply to light waves.

The result of the Michelson Morley experiment would predict that no matter what young Albert's flight speed might be, the light beam would ***seem*** to him to be moving at the speed of light. Now, Michelson hopes that this isn't really so, and waits for an illusion to surface. But no illusion is found.

The message is that light ***seems*** to be travelling at the "speed of light," 2.998×10^8 m/s, no matter how fast the observer is moving. But there is no illusion; in the absence of an illusion, one concludes that what light ***seems*** to be doing is what it ***is doing***. The observed speed is not subject to any correction. This spells bad news for the classical velocity addition theorem. And yet it hung on; even Einstein had trouble letting it go.

4.
Einstein

The result of the Michelson Morley experiment, if one believes it, would seem to provide a way out – albeit a most unlikely way out – for Einstein from his self-imposed predicament that made the relativity principle and Maxwell's equations incompatible. Recall that Einstein clearly ruled out any accommodation that would ascribe the Maxwell speed of light to a special frame of reference, such as that of the æther. The only way out would require Maxwell's Laws to actually be the same in all frames of reference, making the value 2.998×10^8 m/s the speed of light that would be measured regardless of the speed of the observer or the speed of the source of the light.

This is what Michelson and Morley's experiment did in fact show.

It violates the classical velocity addition theorem. It is as if a passenger's velocity, sitting still in the front seat of a moving car, is the same when observed in the car as when observed from fixed markers at the side of the road.

Because this and countless other examples that you can imagine have such a powerful intuitive logic behind them, it did not occur even to Michelson and Morley that what seemed to be the result of their experiment could be true. And so they searched for the illusion that was making it *seem* true, when they believed that, clearly, it could not *be* true.

There were candidates for such illusions. One idea proposed that the earth "drags" the æther with it, and thus appears always to be at rest with respect to it. However, if this were to be so, it would be difficult to explain numerous astronomical observations, assuming that the æther far away from earth is immune to earth drag. All of the illusion-candidates that were floated eventually sank, because they predicted effects which experiments could not confirm.

When there are no illusions that can translate one's observations of what ***seems to be*** into what one passionately believes ***is***, then one must consider that it may be time to accept one's observations. Perhaps it is time to adjust theory to fact, rather than continue to try to bend facts to what one believes theory to be.

One would imagine that this is the path by which the Maxwell conundrum would be resolved. But, in fact, historically, the process followed a more circuitous route.

Einstein's summer of 1905

There is an intriguing historical puzzle in the relation of Einstein's development of his theory of relativity to the Michelson-Morley experiment. The experiment was done fully 18 years before Einstein published his first paper on relativity in 1905; the result of the experiment, that the speed of light seemed to be the same in all frames of reference, was a key element of his new theory of relativity; and yet, Einstein did not mention the Michelson-Morley experiment even once in his published work of that time. Historians are not clear whether Einstein knew of the experiment when he wrote his 1905 paper; it is hard to believe that he didn't. Einstein himself years after recollected it both ways, but left little doubt that it was unimportant to the emergence of the theory of relativity in his mind.

This is not to suggest that Einstein operated in a vacuum of experimental fact. He was familiar with evidence obtained by the Dutch astronomer De Sitter on aberrations in observations of double stars. He also was familiar with an experiment of A. Fizeau on the velocity of light in streaming water. Both of these already hinted at the Michelson-Morley result. It is quite possible that Einstein was so convinced of the general truth of Galilean

Relativity, that, had Michelson not done the experiment, Einstein would have fully expected it to turn out as it did. The experiment may have seemed a rather nice touch, such as finding out that indeed the sun does rise every morning, but not necessarily earth shaking.

Einstein had discarded the idea of the æther a long time before. The ardent attachment of most physicists of the time to the æther concept had no evidence to support it – beyond the homily that to ride to town you have to have something to ride on. It just seemed unimaginable that a disturbance such as a light wave could exist if there was nothing to be disturbed. Einstein simply did not suffer such ardent attachment. He thought the æther "unnecessary."

Moreover he was aware of the work of H.A. Lorentz, who had been working with the Michelson-Morley result for some time.

Like many others, Lorentz was in search of the illusion that might make it *seem* as if the speed of light were the same in all frames of reference. No physical model that had been explored was able to survive closer scrutiny. Failing a successful physical model, Lorentz tried a different line of attack. He asked the following question: If there is an illusion that can explain the constancy of the speed of light, then what would a mathematical formulation of that illusion have to look like?

To understand the question, imagine that you look through a device and find that everything on the other side of it looks twice as big as it does without the device. While you might not know about lenses and telescopes, and therefore would have no physical explanation of the illusion, you could conclude that the mathematical effect is multiplication of all distances by two.

This mathematical analysis of the illusion would be written as,

$$d' = 2\,d$$

and would be called a "transformation" of distances (d) to what these distance *seem* to be (d') under the illusion created by the device.

The classical velocity addition theorem, Eq 3.2, is also a transformation. It transforms the velocity, v, that is observed in one frame of reference to the velocity, V, that is observed in a frame of reference with respect to which the original frame is moving with velocity, u.

Lorentz asked what mathematical transformation would create the illusion that, instead of finding the velocity of light that would be expected in a moving laboratory according to the classical velocity addition theorem, one finds that velocity to be, always 2.998×10^8 m/s.

Lorentz succeeded in deriving such a transformation. It consisted of two equations, that, together, transformed the original distance and time intervals of a motion measurement to new distance and time intervals, under the mathematical constraint that if the ratio of distance to time is 'c' in one frame of reference, it will also be 'c' in any other frame of reference (where 'c' is 2.998×10^8 m/s).

We will deal in more detail with the Lorentz transformation in chapter 12, because it is a key element of Einstein's relativity theory.

An intriguing historical aside [1] concerns the fact that Lorentz working in the early 1890's was unaware that his derivation had already been done in 1887 by a little known mathematician named W. Voigt, who noted that equations of the type of Eqs AI.15 and AI.21 in Appendix I of this book retain their mathematical form (read: the laws of physics look the same...) under the transformation we now know as the Lorentz transformation. Not until 1906 was Voigt's work recognized in the context of relativity.

Attempts were made by those working in this area, Poincaré, Lorentz, FitzGerald, and others, to give physical meaning to Lorentz' mathematical construct. These efforts were circumscribed by their authors' adamant adherence to an æther-filled universe. Their failure to consider the possibility that the assumption of

---

[1] A. Pais, *Ibid*, p121

absolute time and fixed space might be false was, in retrospect, fatal to their effort.

The mathematics suggested that if there were changes in the time and distance scales in rapidly moving frames of reference, these could account for the paradoxical constancy of the speed of light. FitzGerald believed that actual "contraction of lengths" occurred in fast moving objects universally, that somehow objects that flew rapidly through the æther became squeezed, but he saw it as a physical squeezing that was not related to the observer's motion. Lorentz maintained that deviations from absolute time were localized, and continued to insist that there is but one "correct" time.[2]

It remained for Einstein to consider that it might be the very nature of time and space that we had wrong in our ancient wisdom. It is this that made Einstein's relativity profoundly different and totally original.

A Ray of Light from the patent office

Quite aware of Lorentz' work, Einstein, from his desk at the patent office, puzzled over these questions. It is not to be thought that because Einstein ultimately found a most elegant, if drastic, resolution to Maxwell's conundrum, that this came easily or quickly to him.

Einstein said later, "... at the time I felt certain of the truth of the Maxwell-Lorentz equations in electrodynamics... This invariance of the velocity of light was, however, in conflict with the rule of addition of velocities we knew of well in mechanics.

"I felt a great difficulty to resolve the question why the cases were in conflict with each other. I had wasted time almost a year

---

[2] A. Pais, *ibid*, Chap 6.

in fruitless considerations, with a hope of some modification of Lorentz' idea, and at the same time I could not but realize that it was a puzzle not easy to solve at all."

In May, 1905, Einstein stopped in to see his friend and colleague from the Bern patent office, Michele Besso, to discuss a question "which was difficult for me to understand... Trying a lot of discussions with him, I could suddenly comprehend the matter... My solution was really for the very concept of time, that is, that time is not absolutely defined..." [3] In June Einstein submitted the paper on relativity.

He had finally reached the point where, "as a result of the physical conceptions of time and space, it became evident that *in reality there is not the least incompatibility between the principle of relativity and the law of the propagation of light*, and that by systematically holding fast to both these laws a logically rigid theory could be arrived at." This came to be the *special theory of relativity* that is the subject of this book.

The way out came from his sudden realization that what created the conflict between the relativity principle and the law of propagation of light was the law of addition of velocities.

Einstein had seen in the Lorentz transformation a description, not of an illusion, but of the truth about time and space. The Lorentz transformation was not, he argued, an explanation of why things seemed different from the way they are, but a description of the behavior of time and space which explains why things are the way they seem.

Because that transformation was created to fit the constancy of the speed of light that seemed to be dictated by Maxwell's equations, the new view of time and space appeared to flow

---

[3] A. Pais, *Ibid*, p139. Also, in the same reference, p166, the following quote from H.A. Lorentz, one of the major contributors to the mathematics of relativity, and a generous champion of Einstein, still eight years later, in a 1913 lecture, "... as far as this lecturer is concerned, he finds a certain satisfaction in older interpretations, according to which the æther possesses at least some substantiality, [and] space and time can be sharply separated..."

inevitably from "systematically holding fast" to those two shiny seashells, as Einstein was determined from early on to do.

Einstein's Theory of Relativity

What then was Einstein's Theory of Relativity?

It was simply this: a declaration that two principles must be systematically held fast as postulates of modern (special) relativity.

Einstein's postulates of special relativity

1. The laws of physics take the same form in all inertial frames.
2. The speed of light in free space has the same value for all observers, regardless of their motion, or the motion of the object that emits the light.

The two principles that had seemed in conflict, the law of the propagation of light, and the principle of classical relativity applied not only to laws of mechanics but to all physical laws, were freed of their incompatibility and both were retained by giving them a status more central to nature's laws than the assumptions of absolute time and fixed space.

It should be understood that Einstein's reasoning, so much of it based on what seemed compelling, natural, and simple, was in no sense a proof of its correctness. For this reason, the idea was proposed in the form of postulates. A ***postulate*** is a proposed basis for drawing inferences and conclusions; the postulate becomes accepted over time when the inferences and conclusions are verified.

Einstein had solved Maxwell's conundrum. But at what cost? It left two important physical theories intact. It preserved the simple Newtonian interpretation of ordinary events involving ordinary speeds; and for circumstances involving speeds close to the speed of light (many times faster than our fastest rockets), there were implications, some of which became evident only later, that were profound and diverse, and in some cases left even Einstein bemused.

Had not physicists, despite the most sophisticated mathematical wizardry, come up short by 1905 in reconciling Galilean Relativity and Maxwell's Equations, Einstein's approach might well have been dismissed as the most unlikely solution to Maxwell's conundrum. As it is, it got mixed reviews. H.A. Lorentz, one of the foremost in the array of brilliant mathematicians working on a solution, applauded Einstein almost from the first. The author of the Lorentz transformation generously credited Einstein with putting a physical reality to the mathematical form. Yet, in 1921, sixteen years after his paper on special relativity, when Einstein was awarded the Nobel Prize, it was given for his work on the photoelectric effect, with no mention of relativity. Relativity was still in 1921 considered debatable and unproven.

There was an understandable reluctance to accept a universe without the anchors of a common time progression, seemingly more complicated than anyone had thought. The need to devote entire books, like this one, to the implications of the loss of absolute time, seemed to fly in the face of Percy Bridgman's assertion about the reduction of the complicated to the simpler. It remained to examine whether, all things considered, this was an addition of complexity, or, possibly, an actual simplification of a condition that only looked simple but hid its complexity behind unsolved conundrums.

The Perverse Simplicity of Maxwell's Law

It was probably a lucky thing that at the time of Maxwell's conundrum people didn't know anything about galaxies, nor very much about electronics, the nature of the chemical bond, and molecular biology.

What they did know, however, should have given them pause enough if they had any thoughts of feeling smug about the simplicity of the æther-filled "classical universe," in which time is absolute, space is fixed, and the classical velocity addition theorem holds sway.

What they knew, and what you also know if you have ever been amused listening to the squeaky voice of a friend who had inhaled helium, is that resonances that occur because of the bouncing of sound waves off the boundaries of a cavity, vary in the frequency of the pitch of the sound produced, in proportion to the speed of the waves.

Sound waves travel in helium nearly three times as fast as in air. And so, any wind instrument, which includes your voice box, played in helium resonates over an octave higher than it would in air.

The generation of electromagnetic waves is done in analogous fashion, by a resonant circuit, usually composed of a capacitive and an inductive element. The electromagnetic waves in a resonant circuit bounce in somewhat more complicated fashion than sound waves in a trumpet, but bounce they do.

Those who are familiar with simple electronic circuits know that the resonant frequency, f, is equal to $g / \sqrt{\epsilon_0 \mu_0}$, where g is a geometric factor related to the shape and size of the capacitor and inductor. With Eq 3.1, this makes the resonant frequency of the electromagnetic waves,

$$f = g\,c \qquad [4.1]$$

If you lived on a moving planet in the "classical universe," the velocity of light that you would have to put in for c in Eq 4.1 is not 2.998×10^8 m/s, nor is it the same in all directions. Because of the vector nature of the expression [3.2], $V = (c + v)$, the transformation of Maxwell's equations between the fixed frame of the æther and the earth's frame, would have to give a direction-dependent velocity of propagation of electromagnetic waves. What happens to the dielectric and permeability constants in that case is not at all clear. The capacitance of a capacitor and the inductance of a solenoidal coil of wire, presumably change as the orientation of the devices is rotated.

Above all, however, the resonant frequency of an electromagnetic wave generator varies with direction. On a planet like ours, whose velocity through the æther is a mere 67,000 mi/hr (assuming the sun is at rest in the æther), the resonant frequency of such a generator would vary by a scant 0.01% if rotated, and between April and October, perhaps not a matter of concern.

If the Milky Way galaxy, on the other hand, were in motion, as some galaxies are, with respect to other galaxies, at half the speed of light, then occupants of a planet in that galaxy could experience quite large resonance shifts with rotation of their instruments. Just as the velocity of a swimmer is different going with, against, and across the current in a river, so would the speed of light and therefore the resonant frequency be different in different directions.

To us with our modern technology, it would raise questions such as the following: Would radios work on moving planets, in moving galaxies? or would they be de-tuned each time they were rotated? would cameras de-focus if turned? would cyclotrons cease to work? Why have none of these difficulties arisen on our planet?

Imagine by analogy that sound were to travel on a medium such as æther, rather than on air. Sound would then penetrate the brass tubes of musical instruments, and in a windy outdoor concert the resonances of the air in the instruments that determine their pitch would be affected by the wind.[4] The pitch of the sound could go up or down a few notes as the instrument is turned with the wind, against the wind, or across the wind (Fig 4.1). One should not have such obstacles to good music in a well-designed universe!

Fig 4.1 If air were like the æther, trumpets would change pitch as they are turned in the wind.

This fantasy is an analogy. The air on which the sound travels does not pass freely through the brass walls of the instruments, as the æther would. We can be grateful for that fact, else we would expect our musical instruments to encounter these problems when played in a breeze.

Far more cataclysmic than the penetration of æther winds into electronic equipment, would be the effect on all the myriad chemical and biochemical mechanisms that are electrical resonances at heart.

These consequences of a (hypothetical) classical universe are not incidental matters at all. The problems inherent in a fixed-frame universe are staggering. Might not abandoning absolute time and fixed space in favor of a universal and constant velocity of light and electromagnetic waves ultimately make it much easier to design a universe that works, not only on the cosmic scale, but also in respect to these every-day manifestations!

A co-worker at the Institute for Advanced Study at Princeton, Ernst Straus, remembers that Einstein would sometimes be attracted to an idea because it seemed to him that no sensible architect designing a universe would do it any other way. Straus has recounted a time when Einstein noted a particularly attractive

---

[4] Forget, for the sake of the example, that the porousness of the instrument would make it difficult to establish these resonances.

feature of some new approach, and exulted, *"This is so simple God could not have passed it up."*[5]

Putting yourself in the place of the hypothetical architect of the universe, is it not conceivable that you would design the world around a relativistic model in order to avoid the potentially catastrophic consequences of a classical model?

---

[5] Banesh Hoffman, *Albert Einstein, Creator and Rebel,* Viking, New York, 1972 p227. Einstein, though a spiritual and religious man, did not believe in a personal, Biblical God, who interferes daily in the world. His reference in this context is to God, the "architect of the universe."

5. Relativity, Look-back time, and other confusions

Some of the consequences of "holding fast" to the twin principles of relativity and Maxwell's law of the propagation of light are perplexing, astounding, in some cases hard to understand, at other times easy to confuse. And so, herewith a chapter to familiarize you with some of the traps and ways to avoid confusion.

There is no pretense that the assertions made in this chapter are fully explained or justified. All that will come later. Consider this chapter a guidebook to a foreign country, with hints about the nature of the terrain, warnings of danger, guidelines to behavior. In the table of contents you have already been given a map of the points of interest that you will visit. Not until you actually set foot in the new land will you begin to learn the language and understand fully the meaning of what you will see.

Relativistic effects are real.

Relativistic effects do not *seem* to be, they *are*, in the sense of the previous chapters. Strange as some of these effects are, there are no corrections based on classical (as opposed to relativistic) principles that can be applied to "explain" relativistic effects and reveal a different reality behind them.

Indeed, you may find yourself searching earnestly for such explanations. The consequences of abandoning fixed space and absolute time in favor of the universal value of the speed of light and the broadened principle of relativity, can be in some cases bizarre and unbelievable. Because the phenomena occur in situations involving such great speeds that we have no intuition for them built on personal experience, we have few touchstones from which to make reality checks. We have to rely on our ability to apply the facts of relativity correctly.

The sometimes unbelievable outcomes of relativity are, of course, not random. We cite two quite wild examples here without explanation only to show how far from the familiar and expected they can go.

Imagine, for example, that the captain of a space ship solicits you to take a journey to the Andromeda galaxy. Even this nearest neighbor of our own Milky Way galaxy is incredibly far away, 12,000,000,000,000,000,000 miles away – twelve billion billion miles. It takes a light beam 2 million years to make this journey. The captain assures you that, travelling not even as fast as that light beam, you can make this trip (and back) in your lifetime.

This remarkable achievement, you are told, is possible, according to the theory of relativity, through a slowing of the progression of time resulting from a speed close to that of the light beam, or from a different point of view, through a foreshortening of the distance resulting from the great speed.

Or, consider this unexpected outcome for high speed electrons: In classical physics the Kinetic Energy of an object is proportional to the square of its speed ($K = \frac{1}{2}mv^2$), so we expect to be able to double the velocity of an object by quadrupling its Kinetic Energy. However, if we quadruple the energy of a 10 Mev electron (an electron whose kinetic energy is 10 million electron volts), increasing its energy to 40 Mev, the electron's speed does not double, but increases by only about one tenth of one percent. Moreover, the next quadrupling, to 160 Mev, has an even smaller effect on its speed.

The casual voyeur relishes the bizarre aspect of relativity, comforted that it is distant and unlikely to touch him. He simultaneously believes it because it has status and disbelieves it because it makes no sense. He idolizes the inventor of all this, a distant man named Albert Einstein, whose hair is unkempt and whose brain is beyond comprehension, and simultaneously sees in Einstein's face the ultimate image of the mad scientist.

Abraham Pais perceptively describes the public's fascination with relativity and the mystique that surrounds it. That fascination, as well as "the universal appeal of the man who created all this novelty," he says, arise out of "the nearness to mystery, unaccompanied by understanding." [1]

Fig 5.1 Fast bikes: relativistic length contraction at 600 million mph. "... just an adventure in the bizarre?"

Many popular conceptions of relativity reside on the fuzzy border between fantasy and reality, causing some who don't care to give the matter deeper thought, to conclude that the ways of science, and of scientists, are strange, unknowable, and not necessarily to be trusted. What a sad outcome!

The image of time and length distorted, as if seen through lenses that fool the eye, is the unfortunate result of displaying the phenomena of relativity in isolation from the science of which it is a part. In the absence of understanding, relativity is no more than an adventure in the bizarre, a fast-food intellectual meal, strongly flavored but devoid of nutrition.

It is not surprising that we have little in our intuition to help us form mental pictures of relativity's more surprising occurrences. Because relativity is outside the realm of our experience, we have no inner guide to what is believable and what is not.

The yearning to "understand" relativity, however, can not be gratified by suggestions that there are non-relativistic principles

---

[1] A. Pais, *Einstein and the Press*, Physics Today **47** **#8** Aug 1994 p30

that can explain relativistic phenomena. Only by invoking the theory of relativity itself can we know why relativistic phenomena occur.

An elephant is not a large dog

Relativistic effects can not be explained away as just extreme examples of non-relativistic effects. Just as an elephant is not a very large dog, so relativity does not consist of ordinary distortions magnified by speed.

Nor is a miniature elephant a dog: It is tempting to think that if we had eyes that could slow down the images of things that fly by at relativistic speeds, making our minds capable of photographing those images, the relativistic phenomena would become nothing more than exaggerated versions of what we see in slowly moving objects. This is decidedly not so.

Relativistic effects are not simply extensions of effects in the slower moving world; they have to be understood on their own terms.

At the same time, while it is true that relativistic effects are outside our ordinary experience, relativity is not mysterious.

The postulates of relativity are simple and believable, and can be verified by experiments that almost anyone can understand.

Some of what you will encounter in undertaking to understand relativity may, and in fact, should, cause your intuition to rebel. The purpose of your intuition, built up from years of experience, is, after all, to provide you with a built-in auto-pilot that guides your response to familiar circumstances and raises alarm at discrepant events. It would be unfortunate indeed if your intuition failed to express alarm when you see time and place about to shift from under your clocks and maps. Indulge your healthy skepticism about these new and unfamiliar rules. Keep asking the questions that children are taught not to ask. It is through them that you can eventually reach a comprehension of relativity.

Relativity, Look-back Time, and Optical Illusion

One illusion that is frequently (but wrongly) associated with relativistic effects, is the appearance of distortion in the shapes and the relative location of objects due to the time delay involved in the travel of light between the viewed object and the observer. This is an illusion because it is due to a well understood, non-relativistic effect, that can be corrected for.

Perhaps you know that it takes eight minutes for light from the sun to reach us. Therefore we see the sun always as it was eight minutes earlier. This eight-minute delay in the arrival of the image of the sun for us to see, is called "Look-back time." We see galaxies with look-back times of billions of years. We see them as they were billions of years ago.

Look-back time is not a relativistic effect. It can produce a distorted view of an object if different parts of the object have different look-back times. For example, at a particular season, Mars and Earth may be at opposite sides of the sun in their orbits, which would put Mars more than twice as far from earth as the sun. Mars would be seen from earth as it was some 18 minutes earlier, the sun as it was eight minutes earlier. The configuration of sun and Mars will appear slightly distorted because of this. The effects of look-back time can be understood completely from classical physics, and can be mathematically corrected for. This is geometry, not relativity.

In the same way, any rapidly moving object may appear distorted because the light from different parts of it will experience different delays before its arrival at the eye (or camera). Such distortions are most pronounced when the object moves at nearly the speed of light, the same circumstance under which relativistic effects occur. It is easy to confuse relativity with the effects of look-back time. (A good criterion by which to sort out these distinct effects is given in a footnote[2].)

> *A Rule of Thumb*
>
> 1. If it can be explained in classical terms, it isn't relativity.
> 2. Relativistic effects can not be understood in classical terms.

The Clock Tower and the Trolley Car

An extreme version of a look-back time effect masquerading as relativity is contained in a film that is still sometimes shown to students as an introduction to relativity. This film shows a trolley car speeding away from a clock tower. When the trolley reaches a speed close to that of light, the time on the clock tower appears to the trolley passengers to cease advancing, since of course their view of the clock is forever formed of light which left the clock at the same time that they did. Light that left the tower later can not ever catch up with the trolley, since the trolley is moving at the same speed as the light.

What is wrong with this film is not that the trolley is travelling impossibly fast (though it is). What is wrong is that the easy and implied interpretation is a look-back time effect, totally non-relativistic. In addition, the truly relativistic effects are ignored; these would lead to the conclusion that, in the view of its passengers, the trolley cannot outrun the light from the tower!

If you remain unconvinced that the simple effects in the trolley example are *not* due to relativity, consider a clock that *chimes* the minutes, and a trolley moving away at the speed of *sound*. The passengers of such a trolley will never hear a chime that leaves the tower later than the trolley did; the sound of such a later chime will never catch up with that trolley that is travelling at the same speed as the sound. Sound moves slowly compared with light, and no relativistic effects are expected in examples involving sound. The "illusion" is the same, can be corrected for in the same manner, and is completely explainable without relativity.[2] These passengers, by the way, can actually outrun the sound from the chimes.

A general caution: When you feel like exclaiming, "Ah, this makes sense. *Now* I understand relativity," be sure that what you think you understand is not, in fact, an illusion due to ordinary effects, which has been mistaken for a relativistic effect.

In any example designed to illustrate relativity *correctly*, one first accounts for all the circumstances of observation, such as delays in the arrival of signals. Only after all the "illusions" are accounted for, does one then look for relativistic effects.

What follows in the next three examples is a comparison of relativistic effects and look-back time effects on moving bicycles as seen by a stationary observer.

---

[2] A good test to distinguish relativistic effects from illusions due to delays in the arrival of light from different parts of an object, is to imagine the object illuminated by ultra-sound, in the way that bats "see" objects at night. If the effect is still there when sound, rather than light, is the method of observation, then the effect is not relativistic.

(Note: The speed of the object would have to be reduced by the ratio of the speed of sound to that of light, meaning that an object travelling at 90% of the speed of light would be depicted as "seen by sound" and travelling at 90% of the speed of sound.)

Bicycles: 1. Normal speed

Fig 5.2 Normal bicyclists at ten miles per hour.

Consider the following example that deals with the relativistic shortening of lengths in rapidly moving objects. (The meaning of this statement will be made much more precise later on.) Imagine two bicyclists moving rapidly past a house, observed by someone standing in the foreground, at rest. At moderate speed (10 mi/hr) the bicycles appear undistorted (Fig 5.2).

Bicycles: 2. Relativistic speed

Now imagine that the bicyclists attain a speed comparable to the speed of light, let us say 600 million miles per hour. (This is about 90% of the speed of light.) There will be a relativistic shortening of the lengths of both bicycles by a little more than 50% (Fig 5.3). This is observed even after all effects due to the time it took light from various parts of the bicycles to reach the observer are corrected for.

Fig 5.3 At 600,000,000mi/hr. shortening of bicycles is not an illusion; it is a relativistic effect.

Bicycles: 3. Distortion due to "Look–back Time"

Next translate the problem into one in which ultrasound is used as the observing signal. The bicyclists are assumed to be travelling at 90% of the *speed of sound*, and an image is formed out of the sound waves being received by the observer. (Imagine that the observer is a bat! In their so-called "night vision," bats use their hearing, by which they "see" images of ultrasound signals bounced off objects.) Because the speeds involved here are far below relativistic speeds, there is no relativistic shortening.

Fig 5.4 Bicycles at 670mi/hr 'seen' with ultrasound (note 2). Effect is due to look-back time, not relativity.

The ultrasound, however, takes longer to arrive at the observation point from parts of the bicycles that are farther away. The wheels at the extremes – left and right – are farthest from the viewer, and hence appear as they were at an earlier time, farther back on the left and closer in on the right. (Fig 5.4) The effect is noticeably different from the relativistic shortening of Fig 5.3. The left bicycle appears to be stretched while the right one appears shortened, an effect due to differences in *look-back* time at different portions of the bicycles.

It is possible to correct for this illusion entirely within the rules of classical physics, using the speed of sound and the geometry of the sound paths to calculate look-back times. In Fig 5.4a the ultrasound "view" of the bicycles of Fig 5.4 has been corrected for look-back time effects. No effects remain; speeds are much too slow for relativistic shortening.

Fig 5.4a Same as Fig 5.4, but corrected to eliminate illusion due to look-back time delay of the ultrasound waves.

"The speed of light is the same to all observers, regardless of their motion"

The apparent simplicity of this statement of Einstein's second postulate of relativity can entice us to accept it too easily. That two observers having different motions themselves might agree upon the velocity of a third object does not, after all, seem remarkable.

We mean different things when we say, "We all agree that the speed of light is 3×10^8 m/s," and when we say, "We all agree that the speed of the truck is 75 mi/hr." The first is a statement that 3×10^8 m/s is what we all find when we do the measurement from various moving laboratories. The second is a statement that our measurements agree after being corrected for the illusion created by our motion.

This distinction is so profound, and so important, that two examples in each category are presented.

Example 1. *Classical*

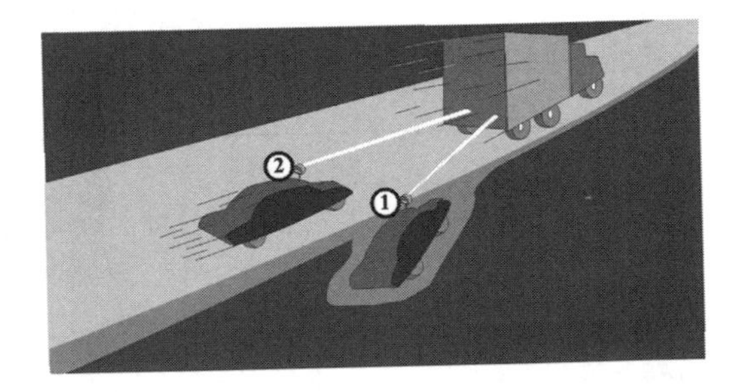

Fig 5.5 Two patrol cars track a speeding truck. Do their instruments read the same speed?

Two police officers corroborate each other's testimony that a particular truck was travelling down the road at 75 mi/hr, even though one officer was making that observation from a patrol car (#1) standing at the side of the road while the other officer was in a moving patrol car (#2), pursuing the truck at a speed of 65 mi/hr, and their police instruments gave quite different readings of the truck's motion.

The officers are said to agree, in spite of the fact that the instruments in their cars gave quite different readings of the truck's speed. We accept the fact that the readings of the instrument are measurements of the speed of the truck ***relative*** to the patrol cars. A correction based on the classical velocity addition theorem [Eq 3.2] causes both officers to conclude that the truck was travelling at 75mi/hr, *with respect to the fixed frame of the roadside.*

Example 2. *Classical*

The transmission of sound waves is a classical wave propagation process, which does not obey the postulate that applies to the propagation of light. It would be correct to say that "Sound always travels at 330m/sec through the air." Because it is understood that sound travels as a disturbance of air, the statement implies that the stated speed applies to the speed of propagation with respect to the air; if the air moves, or if we are in motion against the air, then the speed of the sound as we observe it is the total of the velocity of the air and the velocity of the sound in still air.

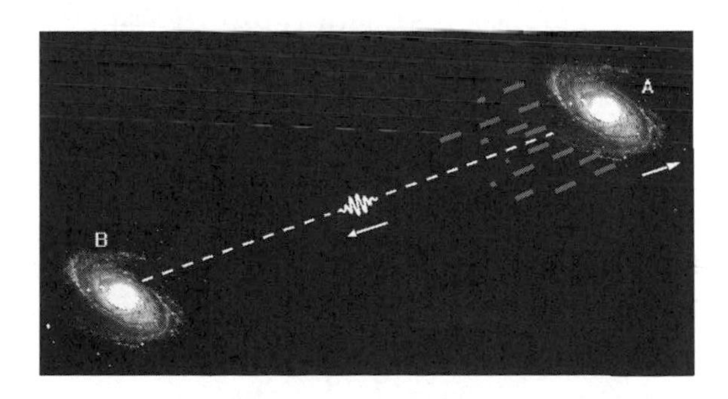

Fig 5.6 The light pulse travels at the same speed relative to observers in both the at-rest and the moving galaxies.

Example 3. *Relativistic*

Imagine, in this example, that a pulse of light is sent from another galaxy, A, that is moving away from our galaxy, B, at great speed. The pulse is received in an observatory in our galaxy. (Fig 5.6). An observer in the galaxy, A, measures the pulse leaving A at a speed of *c*. Even though the light source, A, is moving away from us and our observatory, B, the pulse will be approaching us at the speed *c*. Our naive, classical expectation is that the departure of the pulse at a velocity *c* from the receding source galaxy, A, would have the effect of slowing the pulse relative to us (in B). This is incorrect. There is no classical velocity addition correction to the velocity seen by B that results from the movement of A.

We did not say the light pulse "seems" to us to be approaching us at the speed '*c*.' It really "does" approach us at the speed '*c*.' Just as it really "does" leave the source galaxy, A, at a speed of *c*, as viewed from A. The measurements are not subject to illusions. They are both real.

Fig 5.7 Each observer, the one at rest and the one on the rocket, sees the light beam approaching him at speed *c*.

Example 4. ***Relativistic***

Suppose you are speeding past me in a rocket ship with a velocity of half the speed of light and there is a light beam approaching us (Fig. 5.7). If that light beam is approaching *me* with the speed of light, *c*, one would expect it to be approaching *you* at a speed of 1½ *c*. Yet, measuring instruments in both my laboratory and yours record the speed of that same light beam to be 3×10^8 m/sec.

If it were only so easy!

These examples may have led you to believe that there are just the two cases: (1) ***Light pulses***, which travel with respect to every observer at a speed equal to '*c*.' and (2) ***All other objects***, which obey the classical theorem of velocity addition (Eq 3.2).

That is not the case. It is true that when objects travel at velocities that are small compared to the speed of light, which includes just about all things that we would normally observe in motion, Eq 3.2 is a very good approximation. (In fact the speed even of climbers on escalators and truck drivers clocked by traffic police in different police cars are affected by relativity, very very slightly. But that is not the point here.)

When velocities are very great, but the moving objects are not light wavelets, and do not travel as fast as light, what happens is different from both slower motions and from the motion of light itself. Suppose that in the picture of Fig 5.7, I am approached by a projectile moving toward *me* (I am at rest) at a speed of ¼*c*. *You* (moving toward the projectile at ½*c*) might expect to find, based on the classical velocity addition theorem, that the projectile is approaching at a speed of ¾*c*, but in fact that is *not* what you would find. Nor would you find that it approaches you at the same speed with which it approaches me (¼*c*). The speed *you* would measure would be 0.667*c*. (You will learn in chapter 13 how to do this problem correctly.)

The main subject of the remainder of this book is the impact of Einstein's theory on observations of objects that move very fast, but not as fast as light.

Problem 5.1 – *a problem for you to try!*

The concepts you have seen applied in the examples can seem very abstract. To make them more tangible, let us pose a problem, which, as you will see, is not as simple as it might seem. You will learn how to solve it correctly, later. At this point as you try to solve the problem, you will find that conventional methods give two conflicting results, which can't both be correct.

The boat problem

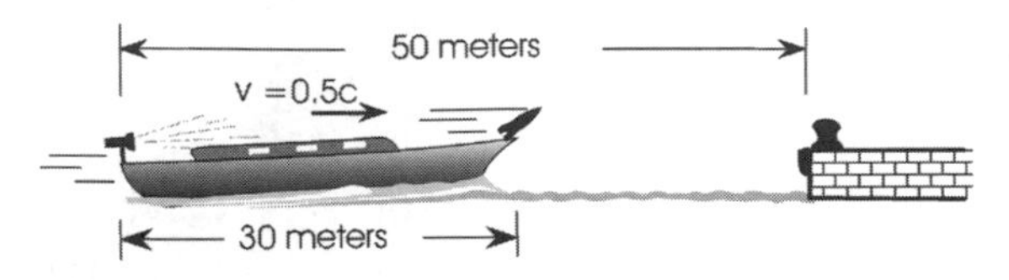

Fig 5.8 Will the passengers survive?

A boat is travelling with a velocity of half the speed of light (0.5*c*) toward a dock. At the stern (back end) of the boat is a single-flash light. At the bow (front end) is a bomb with a light-activated trigger. The boat is 30 meters long.

At the instant when the flashlight end of the boat is 50 meters from the dock, the flashlight sends a pulse of light toward the light-activated trigger at the front of the boat.

There are two possible outcomes: *If the boat hits the dock before the bomb is detonated*, the bomb will be knocked away by the collision with the dock, and will not detonate. The boat and the crew will be shaken up, but will survive. *If the light pulse arrives at the detonator before the boat hits the dock*, the bomb will explode, sinking the boat and killing the crew.

Of the possible outcomes, one, and only one, will occur. Every observer will see the outcome that occurs; *no* observer will see the outcome that does not occur. In other words, whether the bomb explodes and the crew is killed can not depend on the frame of reference in which the observations are made. An *event* either does or does not occur. Which do you think will occur?

Conventional methods will produce different answers, depending on whether you believe the light pulse must travel 30 meters (the length of the boat) or nearly 50 meters (the distance the flashlight is away from the dock) to reach the trigger before the trigger reaches the dock. By analogy with sound waves, you may be searching for the "medium" upon which the light pulse rides toward the trigger. Is the medium outside of, or a part of the boat?

[*The correct solution to this problem is at the end of Chapter 8*]

6. Clocks and Meter Sticks: What can you Keep and What do you have to Toss? – Some Corollaries

We now have a choice of paths that will take us from the postulates of the Theory of Relativity to the phenomena that are implied in them.

One path is formal: Find a mathematical relation that transforms the time and space variables from one to another of any two inertial[1] frames, in such a way that the velocity of light is invariant. We will talk about transformations later; suffice it to say at this point that the transformation that does this (while leaving other things sensibly defined) is the Lorentz transformation. It replaces, among other things, Eq 3.2 for transforming velocities. From the Lorentz transformation, one can deduce many of the consequences of relativity. This is the more abstract way to go.

We will choose a path that is more intuitive and visual: we will stay with the postulates themselves as the basis for deducing consequences.

---

[1] "Inertial" frames of reference are those in which the Law of Inertia (Newton's First Law of Motion) holds true. They are frames of reference that are either stationary or move at constant velocity, that is, they are not accelerated. With respect to what? Well, with respect to other inertial frames. That means you have to start with one in which you verify experimentally that objects left alone (not acted on by any net force) continue to move at constant velocity (this is Newton's First Law) . Then other inertial frames can be defined in relation to that one. In Special Relativity "moving frames" always refers to "moving inertial frames."

The consequences of Einstein's choice to retain the law of propagation of light as well as Galilean Relativity involve certain phenomena that can be said to be "predicted" by the Theory of Relativity. These phenomena are not merely interesting characteristics of the relativistic world that we live in; they are also the predictions by which the Theory of Relativity has been validated over the past century.

The greater meaning of '*c*'

We will find that the speed of light, 2.998×10^8 m/s, called '*c*', has much wider significance in the phenomena of relativity than simply the rate at which electromagnetic waves go from one place to another.

'*c*' is the name of a constant that appears in the wave propagation solution of Maxwell's equations involving dielectric and magnetic properties of empty space: ($c = 1/\sqrt{\epsilon_0 \mu_0}$). Its numerical value, of course, depends on the units that are used, and has no significance in itself. If seconds and meters are used as the units of time and distance, the value of '*c*' is 3×10^8 m/sec. Units can be chosen[2] that make the numerical value of "*c*" equal to one, which has certain computational advantages.

That a velocity appears as a constant in a fundamental equation of physics suggests that this constant is more than just a measure of how fast something travels. How *wasteful* it would be to expend a fundamental physical constant simply to establish the speed, let us say, of a swimmer, or of the earth, or of electrons, or of light! Nature is frugal, and would not countenance such extravagance.

Such a point of view led Einstein to approach the puzzle of Maxwell's equations with the suggestion that the most profound

[2] If 3×10^8 meters is defined as 1 absolute distance unit (1 "adu"), the speed of light is 1 adu/sec

meaning of this universal physical constant with units of velocity lies not in telling us how fast light travels, but in telling us how nature regards the relation between time and space.

A daring, and somewhat preposterous idea!

Are our clocks and meter sticks now obsolete?

If time and space are related, if a minute is not always a minute and a meter is not always a meter, then what measurements can we still make, and how? Certainly the new way of looking at space and time is made difficult by our long and exclusive experience with meter sticks and clocks, which we have always counted on to give us exact and reliable measurements, independent of each other.

Certainly the use of clocks and meter sticks will require some adaptation to the new rules. These we will discover in the next chapters. Meanwhile, consider some illustrations.

A ball is thrown in the air. We use an ordinary clock to measure how long the ball is in the air. The clock gives the answer: 21.452 seconds, let us say. We can assume that this measurement is *accurate*. That does not mean, however, that it is *uniquely correct* – correct without qualification.

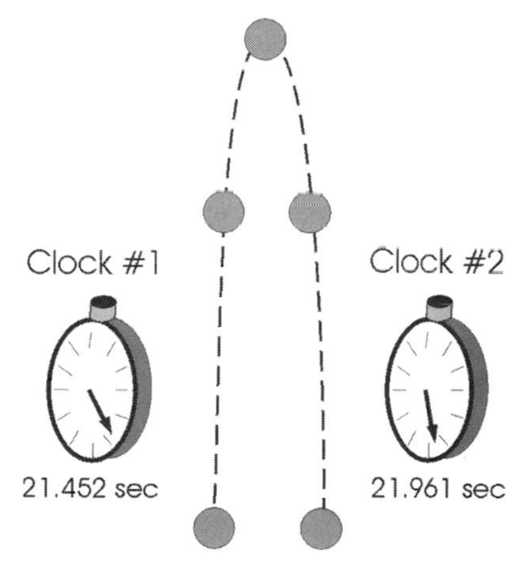

Fig 6.1 Both clocks are right! One clock is in a moving frame of reference.

Time does not "flow equably without relation to anything external," and an equally *accurate* clock, timing the same motion of the same ball but carried by an observer in a different frame of reference, may have measured 21.961 seconds, a result which, under the conditions of that measurement, is *equally correct*.

Consider another example: Suppose galaxy "Q" is said to be 10,000,000 light years away from us. This is a way of saying that its distance from us is such that a pulse of light would take 10,000,000 years to reach us. Given careful interpretation, such a statement has meaning, but saying that "the pulse of light which arrives here today left the galaxy 10,000,000 years ago," is a statement that would be challenged by the inhabitants of the galaxy, even assuming they used clocks identical to ours.

Even more amazing is the time perspective that the photons (the tiny wave packets of light) contained in that pulse of light have of their own journey here. By the photons' own clocks, no time has elapsed at all during the trip that by our reckoning took 10,000,000 years. In the experience of a photon, its emission from the surface of the star and its absorption by a grain of silver bromide in the emulsion of a photographic film in the rear of an earth telescope 10,000,000 light years away, is a single and simultaneous event.

How, you might ask, can a photon move this enormous distance in zero time? It certainly helps that, from the point of view of the photon, the *distance* as well reduces to zero, notwithstanding all the meter sticks we could place end-to-end between that galaxy and ours. More on that later.

With our old ideas of time and space gone, What is left?

When a clock isn't a clock, and a billion billion meter sticks laid end to end can have zero length in some other frame of reference, what is there that endures? What ground remains for us upon which to rebuild our world view?

Many things that have always seemed reasonable and acceptable are apparently incorrect; many rules that we have accepted as obvious apparently do not describe the way the world really is.

Our trust in our intuition is understandably shaken. How can we decide what to discard and what to keep?

Many of our most fundamental assumptions about the world around us are based on what we call our intuition. Our intuition for physical things is amazingly good. Our early experiences have adapted us well to the task of manipulating the physical world around us day in and day out. It is only when we try to deal with situations that are very far from any experience we have ever had, that our intuition abandons us.

It is some comfort to recognize our intuitions for what they are, to realize that they are mental images which seem right because they are garnered from experience, that it is not our mental capacity but our limited experience that is lacking. If in some part our intuitions must be torn down, they can be rebuilt by exposure to new experience.

Very little rebuilding of intuition is required to enable us to accept classical physics, because classical physics is based on observations not too different from those we make in our every day life. Our intuition is not outraged to learn that if we travel 200 miles at an average speed of 50 miles per hour, we will be on the road about *four hours*. The equation from which this result is calculated, though not much simpler than the equations of relativity, is more easily learned, accepted, and used because it gives expected answers. Relativity appears more difficult largely because it gives answers that we don't expect.

In relativity we have to reckon with rules that say that there may be an observer in a different (inertial) frame of reference for whom the duration of our trip would be *five hours*, and that there might be another observer in another frame who perceives the distance we have travelled to be only *100 miles*. These are not mistakes, nor are they illusions. For each of those observers, their results are the outcomes of properly done measurements with accurate instruments and with all optical illusions properly taken into account and compensated.

There is simply nothing in our experience that can fit such facts into a comfortable mental image. That does not make those phenomena impossible; it means that for our intuition to be able to accept them we must become sufficiently accustomed to the conditions under which they occur, so that *there will be mental images* into which these new events can fit.

To do this requires a rebuilding of our intuition *in the presence of experiences in which relativity makes its presence felt.* Since the circumstances under which we might experience relativistic phenomena are not readily present in our lives, we have to rebuild our intuition in the presence of *circumstances that are described*, and that become part of our experience through our imagination.

COROLLARIES: What is still OK to do the old way

Before we can assemble a body of relativistic "thought experiences," it will be helpful to make quite explicit what we believe remains of our familiar strategies from classical physics. By and large, these are natural and obvious guides that are commonly thrown in without comment.

Because no longer can we assume that what seems obvious is true, we propose to list these as "corollaries" of the new rules, and refer to them later on when we use them.

Agreed that the fact that they make perfectly good sense is scant justification. Ultimately, the justification of these corollaries is the same as the justification for the postulates themselves: after one hundred years of verification, they have never failed to describe the world correctly.

Corollaries

1. **All measurements that are made within one frame of reference are still valid.** *Relativity is about differences that occur when the point of observation changes from one frame of reference to another.*

1a. You can still measure the distance between two points in space using a meter stick.

You can still take a steel or platinum rod of a certain length and call that length, by definition "one meter." By comparing other rods to that standard rod, you can make "meter sticks."

If you can lay a number of meter sticks end to end on the ground in a frame of reference in which those meter sticks are at rest (imagine that they can be nailed down, so that even if you move to place them, once they are placed they remain fixed to the frame of reference) then the distance that you read with these meter sticks is a unique and so-called "proper" distance between two points.

1b. You can still measure time using an ordinary clock.

You can still make a clock using any uniformly repetitive motion, such as the swing of a pendulum or the vibration of a quartz crystal.

As long as the clock remains in one place in a frame of reference, you can define "one second" as the time elapsed during some number of vibrations of your clock.

1c. You can synchronize clocks in the same frame of reference.

Two clocks at rest at separate locations in the same frame of reference can be synchronized. One way to accomplish this is to start them with an optical trigger on each clock. If the triggers are set off by two flashes of light sent simultaneously from a point equidistant from the two clocks, the travel-time delay of the light flashes before they reach the two clocks is the same. The flashes will therefore reach the clocks at the same time, causing both clocks to start at the same instant. Once started, such clocks will remain synchronized, as long as they remain fixed (at rest) in the same frame of reference.

Several clocks can also be synchronized by setting each one to a starting time that takes into account the travel-time of the light flash to that clock.

1d. You can measure the speed of light (c).

Having defined a meter in a particular frame of reference, you can now lay out a chosen distance by laying a sufficient number of meter sticks end to end, from point A to point B.

Having defined a second in that same frame of reference, you may now clock the passage of a pulse of light from A to B. You can use clocks at A and B that have been previously synchronized, or you can do the experiment with a single clock by reflecting a pulse off a mirror at B back to A. If there are N meter sticks, the total distance travelled by the reflected light pulse will be 2N meters.

The velocity of light is now equal to the total distance travelled by the light pulse divided by the number of seconds the clock ticked off.[3]

[3] The definition of the units of time and distance have evolved over the years. The second, which used to be defined in terms of the length of the day, is at present defined as the time of exactly 9,192,631,770 vibrations of a particular mode of the natural vibrations (in the microwave region) of a cesium-133 atom. The device that keeps our standard of time synchronized with this definition is called a cesium clock. The meter, which used to be defined as the distance between two marks on a platinum rod kept in a vault near Paris, is now defined as the distance traveled by light in vacuum during a time of exactly 1/299,792,458 of a second. This means that the speed of light is *exactly* 2.99792458×10^8 meters per second. If in the future there is an increase in the precision of measurement of the speed of light, that number will not change, but the length of the defined meter will change.

2. 'Events' are physical realities

An event is something that occurs at a particular time and place (actually nothing has to occur – an event is defined by a time and a place). The time and location of an event may be described by numbers that would be different in different reference frames. But the event is a physical reality quite apart from the numbers that you or any other observer may use to describe it.

An example of an event is the explosion (or the failure to explode) of the bomb in the bow of the ship in problem 5.1. That event is a physical reality, independent of its measurement by an observer.

*The sending of the light pulse described in (**1d.**) from **A** and its arrival at **B** is a pair of events. These two events can be observed in the frame of reference in which **A** is fixed, or in a frame that is moving with respect to the first. The time interval between the events and the distance between them may be different to different observers. We will discover later that there are some properties of the pair of events that are the same to all observers.*

3. Displacements (distances) are frame invariant if there is no relative motion of the frames along the direction of the displacement.

The relative motion of two frames affects only that component of the displacement that is along the direction of the relative motion of the frames.

*The **length** of a moving airplane is different to someone moving with the plane (a passenger) and to someone on the ground. The **width** of the plane is not different to those two observers, because that dimension is perpendicular to the line of motion.*

4. Frame velocities are relative and equal in magnitude.

If the car next to yours at the traffic light appears to be moving slowly forward, your only clue that the other car is moving forward and that yours is not drifting backward, is the configuration of "fixed" objects, such as buildings, the traffic light, etc. In the absence of such reference frame markers, the only thing you can say is that there is relative motion between you and the car next to you.

Out in space, where there are no "fixed" objects, being able to assign the motion to either your car or to the other car would violate the first postulate of relativity. If it were possible to assign more of the relative motion to one car or the other, it would be possible to determine that one frame is "more nearly at rest" than other frames, implying that the laws of physics are special in that frame – a violation of the first postulate.

In sum, there is no at-rest frame nor can an absolute velocity be determined for any object or any frame. All frame velocities are relative to some other frame, and the velocities of two frames of reference relative to each other are mutual and equal in magnitude.

These corollaries form a body of reasonable extensions that will be invoked along with Einstein's postulates of relativity to derive some of the more significant consequences of those postulates. Without these corollaries, the question, "How do you know that you can do that?" would continually pop up and remain without adequate answers.

The Phenomena of Relativity

7.
Time

Our sense of time as an absolute is deeply rooted in experience, even more so than is the case with lengths and velocities. To understand relativistic time requires that our intuition undergo the most profound change in viewpoint.

Once we have a thorough understanding of time, some of the other phenomena of relativity will fall into place with comparative ease. We will have got accustomed, at any rate, to the kind of real life situations in which relativistic effects are manifested.

When we refer to a "time" measurement, we almost always mean a time *interval*. (Rarely does it matter what time it *is*.) A time interval is the time elapsed *between two events*. The symbol, t, is used for an *interval* of time, although Δt is implied.

We will find, surprisingly, that there are no unique answers to the questions, "How much time has elapsed between these two events?" or "How long ago did this happen?"

Some Remarks on "Observers" and "Clocks"

The two words, "Observer," and "Clock," as used by specialists in relativity, are idealizations, but the novice is not always informed of that.

Time measurements are said to be made by "observers" using "clocks," in "the observer's frame of reference."

The emphasis on "the observer's frame of reference" leads many a newcomer to speculate that the effects in relativity are somehow related to the extraordinary difficulty of "observing" things that move at relativistic speeds – speeds comparable with the speed of light.

Many thought experiments in relativity involve pulses of light, and this just adds to the suspicion that relativity may be an artifact resulting from failure to take into account such things as look-back time. (It takes eight minutes for light to reach us from the sun; 2 nanoseconds for our reflection to reach us from the bathroom mirror.)

An "observer" in this context is not necessarily a human, nor even a specific instrument, but an idealized device for obtaining experimental data, fully able to take into account any such artifacts, and to correct for them. These idealized observers, you will find, are capable of the most remarkable feats of observation; rarely is it important that these feats can actually be accomplished.

If there is a delay in observing an event at a distance, then one can imagine a second observer placed close to the event, equipped with a clock appropriately synchronized with the first observer's clock. The observer close to the event can record the time from his clock when the event occurs, and at leisure transmit this information to the other observer. Clumsy though this may be, it makes it possible to eliminate or correct for any errors in time measurement due to optical illusions or transmission delays.

Any task of observation that can be conceived, *in principle*, is within the scope of capability of the idealized observers in this game.

Clocks

An "observer," using a "clock," measures the time between two events, e_1 and e_2.

A clock measures the time between e_1 and e_2 by counting elemental time intervals, called "ticks." A clock consists of some mechanism that produces the "ticks" and another mechanism that counts them and then displays the count in some fashion, either by the movement of a clock hand, or by a digital display.

Usually clocks do not display the raw data (the tick count.) Most clocks display a time interval in seconds, calculated by multiplying the tick count (n) by the elemental tick time (τ): $t = n \times \tau$.

It is usually assumed that the tick time is a unique characteristic of the tick mechanism. For a quartz watch, for example, the tick time is determined by the vibration frequency of the quartz crystal. If the quartz crystal of a particular watch has a vibration frequency of 1.0×10^8 Hz, then the tick time is 1.0×10^{-8} sec. If between two events the clock counts 5×10^7 ticks, then the readout will indicate a time interval of 0.500 seconds.

Having abandoned absolute time, we expect, however, that the time between two events may depend on the frame of reference in which it is measured. Does a clock tell us this?

For ordinary clocks the answer is No. The clock is wrong when observed from a moving frame. Where does it go wrong? It doesn't go wrong in the tick count, n. A tick is an event. There can be no frame dependence in the *counting* of events. If between two events e_1 and e_2 there are 5×10^7 vibrations of a quartz crystal, that number of elemental events is the same for all observers, in all frames of reference.

Where clocks go wrong is in the duration, τ, of the tick time used to calculate the time interval. The new twist introduced by relativity is that the elemental time interval, the "tick" time, τ, may depend on the frame of reference of the observer.

What the Timex quartz clock can't do

If the "tick" time, τ, depends on the frame of the observer, then the *quartz clock readout* will be wrong when observed by an observer who is in motion with respect to the clock. (Actually a team of two observers is required; as we will see, a moving observer can not observe a stationary clock at two different times.) The quartz clock correctly counts n, the number of ticks between e_1 and e_2, but the readout is wrong because the *calculator* part of the clock, which multiplies n by τ uses the "factory" value, τ_o, determined at the clock factory with the clock at rest, a value that is incorrect in the frame of the moving observer.

But how are we to determine how the tick time changes when the ticks are observed from a moving frame?

In the postulates of relativity, the assumption that the tick time determined at the clock factory is an absolute quantity has given way to the fact of the constancy of the speed of light. We need to find a clock whose tick mechanism allows us to use that postulate of relativity to determine how the tick time depends on the frame of reference of the observer.

The Timex clock gives us no way of doing that. In that clock, one second is the time of 1×10^8 vibrations of its crystal, when it is at rest. We have no theory for calculating the effect of observer velocity on the tick time of a quartz crystal.

On the other hand, one second is *always* how long it takes a light pulse to travel 2.998×10^8 meters. If we could construct a clock whose tick is based on this fact, it could tell us how the tick time changes when the observer is in motion. To do this, one lays out a distance scale in a given reference frame and causes a light pulse to travel along a path in that frame. The light path may include a reflection off a mirror, to return the light pulse to its point of origin, permitting the tick counter to remain at rest in the clock.

Thought Experiments

A "light pulse clock" is not as convenient as a wrist watch, but it is easy to design such a clock. It is even possible to build such a clock, although that is not necessary. The behavior of such a clock can be determined by analysis quite as well as by constructing the clock. Such an analysis is called a "thought experiment." The conclusions from such an analysis will be as valid as the principles it is based on.

The simplest light pulse clock consists of a light source capable of producing a very short pulse of light, and a mirror placed a distance "a" away. The distance that the pulse of light travels to the mirror and back to the initial point is $2a$. Since the light pulse travels with velocity c, the time for the round trip is $2a/c$.

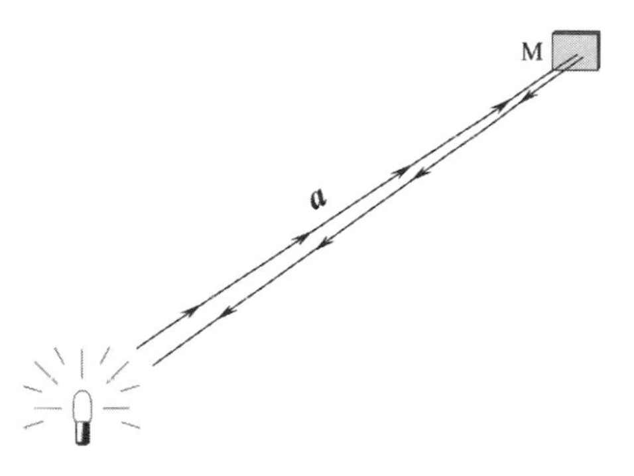

Fig 7.1 A Light-Pulse Clock

$2a/c$ is the "tick time" for this clock in the frame of reference in which the distance a was laid out. A longer time interval can be measured by repeatedly reflecting light pulses over the same path and counting "ticks."

Fig 7.2 A light-pulse clock (left) and a quartz crystal clock (right) keep the same time.

In the frame of reference in which the light pulse clock is at rest, it keeps exactly the same time as a conventional clock that is also at rest. For example, suppose a is 1.5 meters; then the "tick" time of the light pulse clock is 1×10^{-8}sec. A quartz clock with a crystal whose vibration frequency is 1×10^{8} Hertz, has exactly the same "tick" time. If such a quartz clock is placed next to the light pulse clock, the two will "tick" together (Fig 7.2), no matter who observes them.

Suppose that the interval between two events is timed by these two clocks. The read-out of both clocks is correct for an observer at rest with respect to the clocks. The read-out in seconds on both clocks will be *wrong* for a team of *moving* observers because it is based on a tick time that is no longer correct. To the moving observers, both clocks are wrong to the same extent, because time itself, not any particular mechanism for measuring it, is affected by the velocity of the observers' frame of reference (Fig 7.3).

Fig 7.3 The moving observers read the same time measurement on both clocks. Both clocks are wrong.

Clocks are first of all tick counters. They become clocks when "tick time" is built into the mechanism that computes time from the tick count. For the moving observer the tick time τ is longer than the factory tick time; consequently the interval between any two events is longer than that computed by the clock.

The time interval can be calculated correctly for a moving observer by an appropriate re-calculation of the tick time, τ. The re-calculation is the same for all clocks, regardless of their tick mechanism. The unique advantage of the light pulse clock is that there is a theory that can tell us what the required adjustment is.

Fig 7.4 Both clocks read correctly if tick time is adjusted for the moving observers.

The Express Train Stretches Time

Let us see what the theory of the light pulse clock tells us about the effect that the motion of the observers has on the "tick" time. The result can then be applied to clocks with other mechanisms, since we have already established that their time-keeping properties are the same.

For authenticity, if for no other reason, we will use an illustration that Einstein himself used. The illustration is somewhat contrived, but because it deals with familiar objects, it makes for easy visualization; the result is readily generalized. Later we will apply the result to examples from particle physics, where objects are less familiar, but speeds close to the speed of light are common, and relativistic effects are substantial.

In 1900, when Einstein was pondering these questions, trains were a common mode of transportation. Objects inside a railway car can be observed from within the train, as well as from the railway station "platform" outside. A light pulse clock installed in a railway car can be observed by a passenger who is on the train and is at rest with respect to the clock, and it can also, in principle, be observed by persons standing on the platform. Trains do not travel at speeds close to the speed of light (a speed sufficient to take us to the moon in about one second). But, for the sake of this thought experiment, we must fantasize that such a fast train is at our disposal.

Although it is customary to say that it is the train that moves, one can equally well regard oneself at rest if one is on the train, in which case the platform will be moving relative to the train. If the clock is on the train, the latter point of view has the advantage that the observer on the train is at rest with respect to the clock.

Imagine, then, that you are in the train. From that point of view, the platform is in motion. Let us say that the *speed of the train* (as seen from the platform) is v. This is then also the speed of the *platform* as seen *from the train* (Corollary 4, Chapter 6).

Suppose that a passenger (this could be you) has installed a "light pulse clock" at a point A_o on the train The clock is at rest on the train. It consists of a device which can emit a pulse of light, and a mirror placed on the opposite side of the railway car, a distance w away; (w is the width of the railway car). The clock has a sensing device to detect the pulse when it returns after being reflected from the mirror.

The Observer is on the Train

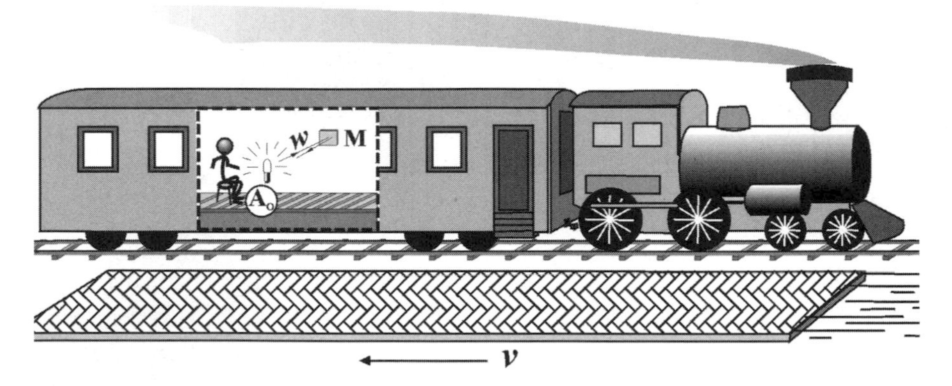

Fig 7.5 To the observer on the train, the train is at rest. In that frame of reference the platform is moving to the left with velocity v.

The passenger, regarding the clock to be at rest, judges that the round trip distance travelled by the light pulse is $2w$, and that the time between its emission and its return is $2w/c$. This interval of time measured in the train frame is the "factory" tick time of the light pulse clock, and will be called τ_o. In the frame in which the clock is at rest, the "tick" time is $2w/c$.

Because the tick mechanism is based on a light pulse, whose speed of travel is known (it is "c") in every frame of reference, it is possible to derive an equation that tells how the tick time, τ, for the observer on the platform, differs from τ_o.

Notation in the two frames of reference

The frame of reference fixed in the train is the one in which the clock is at rest; it will be designated K_0, and variables in that frame (example: τ_0) will carry the subscript ($_0$).[1] The frame of reference fixed on the platform will be designated K; variables in the platform frame will carry no subscript.

The width of the train, measured by laying meter sticks along the floor of the train, is 'w'. Because the width of the car is perpendicular to the motion of the train, this dimension is the same in both frames (Corollary 3, Chap 6); w will need no subscript.

Measuring the "tick" of the light pulse clock from the platform

To determine the "tick" time, τ, of the light pulse clock *as measured from the platform*, one must determine how far the light pulse travels *as seen from the platform*. This requires finding what point on the platform is next to point A_0 on the train when the light pulse is emitted, and what point on the platform is next to A_0 when the pulse returns after reflection from the mirror. But how is one to do this? Recall that it is sufficient to *imagine* doing it!

Suppose that the passenger, at A_0 on the train, with a paint brush deftly marks a spot on the platform, *first* at the instant the light pulse is emitted. Label this spot A (Fig 7.6a). Let the passenger paint another spot on the platform, at B, when the light pulse is received back at A_0 after reflection from the mirror (Fig 7.6b).

---

[1] The K_0 frame may be thought of as the "factory frame," for the clock, since it is the frame in which the tick time is "factory" calibrated with the value, τ_0.

The Observer is on the Platform

Fig 7.6a Event e_1, the emission of a light pulse, seen from the platform. A spot is painted at A.

Fig 7.6b Event e_2, the return of the light pulse after reflection from the mirror, observed at B.

The distance between the spots of paint on the platform can be measured by laying meter sticks from A to B. This distance is a "platform" variable that we name d.

The "tick" time is the time interval between two events: e_1, the emission of the pulse from A_o, (accompanied by the marking of the spot of paint at A); and e_2, the return of the reflected pulse to A_o, (accompanied by the marking of the second spot at B).

Let us examine the path of the light pulse, as seen from above (Fig 7.7). The path as seen from above *in the train* differs from the top view of the path from a fixed point *in the platform frame*. The length of the path taken by the light pulse as seen in the platform frame is equal to $2p$ (see Fig 7.7), which is longer than the path length of $2w$ that one sees in the train frame. The light path is longer in frame K than in frame K_o; consequently the tick time is longer.

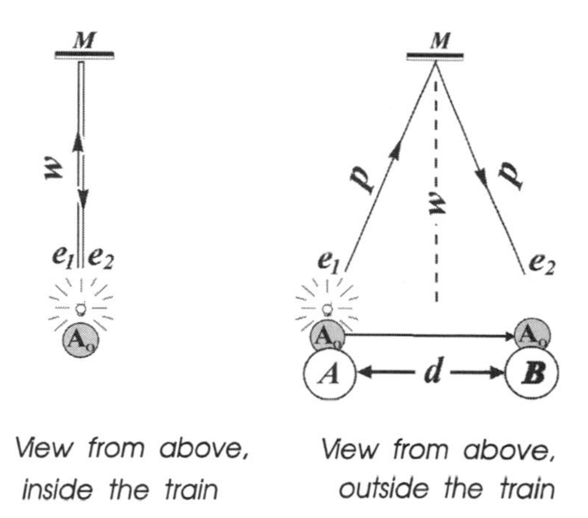

Fig 7.7 The path of the light pulse as seen by train and platform observers.

... which is not to be confused with the "walk across a moving boat" problem

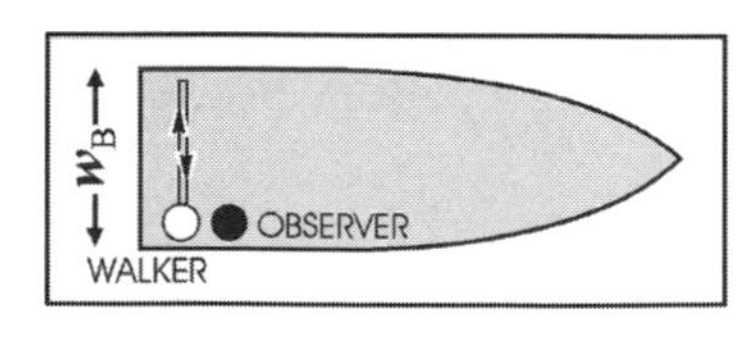

View from the boat

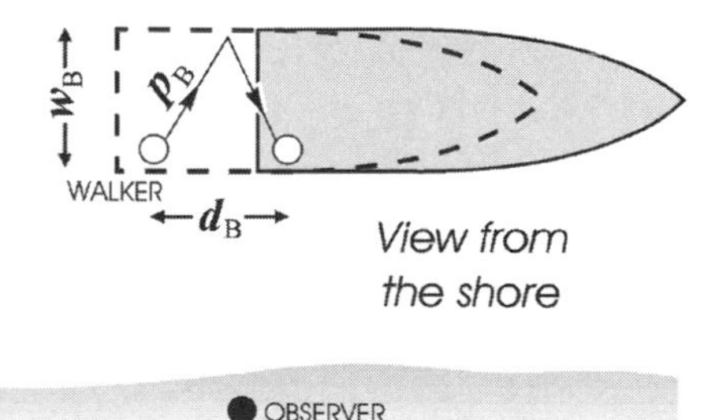

Fig 7.8 The length of the path walked is longer when viewed from the shore.

Fig 7.7 may look deceptively familiar. It looks very much like a diagram that would describe a very simple problem in *non-*

relativistic relative motion (Fig 7.8). Suppose a passenger on a boat that is moving down a river walks across the boat and back. This passenger's walk can be observed by a fellow-passenger on the boat; it can also be observed by a person standing on the shore.

To an observer on the boat, the distance moved by the walker is just twice the width, w_B, of the boat, or $2\,w_B$.

As seen from the shore, the distance walked is $2\,p_B$ (Fig 7.8), a distance clearly longer than $2\,w_B$. The walker traverses a longer distance in the view of the observer on the shore than in the view of a passenger on the boat, just as the light pulse does when observed from the platform. That the distance travelled may depend on the frame of reference is not in itself a relativistic phenomenon.

In the boat example all speeds are small. Relativistic effects are therefore negligible, and the ordinary assumption that time is absolute is justified. We say that the time of the person's walk was of some fixed duration, regardless of by whom observed. If the walker carries a stop watch and determines that it takes him 8.00 seconds to cross the boat and return, then we would not doubt that the observer watching from the shore, and using her own stop watch, would agree that the boat walker takes 8.00 seconds to cross the boat and return, even though she sees a longer path that includes the boat's motion downstream.

The fellow passenger measures the walker's path to be $2w_B$ and calculates his speed to be $(2w_B)/(8.00\text{ sec})$.

The shore observer measures the walker's path to be $2p_B$ and calculates his speed to be $(2p_B)/(8.00\text{ sec})$.

$2p_B$ is greater than $2w_B$, and so the walker's speed measured from the shore is greater than the speed measured by the fellow passenger on the boat. Because in both observations the walk is assumed to take 8.00 sec, we infer that the ***speed is greater*** when the motion of the boat is added to the motion of the walker.

In the boat, the length of the path determines the velocity; in a light-pulse clock the length of the path determines the tick time.

THE DIFFERENCE between the light pulse in the train and the walk across the boat:

In both cases, the distance travelled depends on the frame of reference of the observer.

In the walk across the boat, *time is assumed to be the same* in both frames of reference, so the walker's *velocities are different.*

On the train the *velocity of the light pulse is postulated to be the same* in both frames of reference, so the *times of the motion are different.*

The "factory" tick time, τ_0, of the light pulse clock is $2w/c$. That is in the train frame, where the clock is at rest. Can we determine the expression for τ, the "tick" time observed from the platform?

Refer back to Fig 7.7. In the platform frame the light pulse bounces not from A_0 to M back to A_0, but goes from A to B by reflecting from the mirror M on the other side of the railway car directly opposite the midpoint between A and B .

The path as seen by the platform observer consists of the two segments AM and MB. The *length* of each of these segments is p, which is greater than w. The platform observer sees the light pulse travel a distance $2p$.

How can the same light pulse travel farther in the same amount of time?

How is it possible for a light pulse, whose *speed is the same in all frames*, to travel farther in one frame than in the other, in the same amount of time?

The answer is, of course, that it cannot. If a light pulse has travelled farther at the same speed, *it must have taken longer.*

Understand: this is not *another* pulse that has made the longer and more time-consuming trip. It is the same pulse that has both travelled $2w$ in time $2w/c$ ***and*** travelled $2p$ in time $2p/c$! The time interval between the same two events is $2w/c$ in the train frame, and is $2p/c$ in the platform frame. The difference is not in what the light pulse did, but in the manner in which it is observed.

There are no look-back time effects here. No clocks are used to measure time except the light pulse itself. Only distances are measured; time intervals are inferred from the distances. There is no way to understand this in classical terms. It is a uniquely relativistic effect.

Train Time and Platform Time

We are ready to calculate the time interval τ, measured in the platform frame, and find how it differs from the time interval τ_o measured by the passenger at A_o.

The Pythagorean relation gives us the diagonal distance p from A to M and from M to B in Fig 7.7.

$$p = \sqrt{(w)^2 + (½d)^2} \qquad [7.1]$$

where d is the distance along the platform from A to B, measured in the frame of the platform. The distance d is the distance that the train has moved along the platform in the interval during which the pulse of light has gone across the train and back.

By corollary 3 of Chapter 6, the width w of the railway car is unaffected by the relative velocity of the frames of reference. Consequently,

$$\tau_o = 2w/c \qquad [7.2a]$$

and

$$\tau = 2p/c = 2\sqrt{(w)^2 + (½d)^2}\,/c \qquad [7.2b]$$

The ultimate question is how τ is affected by the velocity of the train, v. To do this we begin by expressing d (a platform quantity) in terms of v and τ.[2] This is easy, since all three of these variables are in platform coordinates,

$$v = d/\tau \quad \text{or} \quad d = v\,\tau \qquad [7.3]$$

Rearranging [7.2a] gives

$$w = \tfrac{1}{2}c\,\tau_o \qquad [7.4]$$

Squaring [7.2b] and using Eqs. [7.3] and [7.4] in the squared terms to eliminate w and d, one obtains,

$$(\tau)^2 = 4\,\{\, \tfrac{1}{4}c^2\tau_o^{\,2} + \tfrac{1}{4}v^2(\tau)^2 \,\} \,/\, c^2 \qquad [7.5]$$

Grouping factors of τ_o and factors of τ, one obtains,

$$(\,1 - v^2/c^2\,)\,(\tau)^2 = \tau_o^{\,2} \qquad [7.6]$$

and, finally,

$$\tau = \tau_o\,\{\,1\,/\,\sqrt{(1-v^2/c^2)}\,\} \qquad [7.7]$$

The factor $\{1/\sqrt{(1-v^2/c^2)}\}$ which appears also in many other relativistic expressions, is given the Greek symbol, γ (gamma).

Because all time intervals are simply multiples of tick times, we can generalize from tick times, τ_o and τ, to time intervals in general, t_o and t. This gives the relation that is referred to as the *time dilation equation*.

---

[2] d_o would be the distance that a fixed point on the platform moves with respect to the train during the interval from e_1 to e_2; it is not equal to d. d_o is neither available nor needed.

The time dilation equation

$$t = \gamma t_0 \qquad [7.8a]$$

$$\text{where } \gamma = 1 / \sqrt{(1 - v^2/c^2)} \qquad [7.8b]$$

The time dilation equation is a special case

The time dilation equation [7.8a] — and the phenomenon called time dilation — is a time interval comparison of a rather specific sort. The equation carries an important restriction, arising from the conditions of the derivation. Time dilation is a special case of the more general relation between time intervals in different frames of reference, the Lorentz transformation.

What makes it a special case is that in one of the frames of reference, the K_0 frame of the railway car, the two events involved occur at the same location. The events are not at the same location in the platform frame.

The time dilation equation is a relation between t and t_0, the time intervals between a pair of events in two different frames, *one of which is a frame in which the events occur at the same location.*

In the time dilation equation, the K_0 frame is that unique frame in which both events occur at the same location.

The observer at rest in the K_0 frame moves along with the location of the events, to be present at both of them.

The K frame may be any other frame.

Proper Time

The message of equation [7.8] is that the time interval between the two events is not unique, but depends on the frame of reference in which it is measured.

There is one unique measurement of this time interval — the time measured in that frame of reference (if such a frame exists) *in which the two events occur at the same location.*

It is not always possible to choose a frame of reference in which two given events occur *in the same location*. For such a frame to exist requires that an observer be able to move fast enough to be present at both of the events.

It is obvious that if two events occur at different locations simultaneously, an observer can not move from one location to the other to be present for both. Even if the events occur shortly after one another, it may still be impossible for an observer to move from one to the other quickly enough (depending, of course, on the distance). The exact conditions when this is so will become clear shortly.

If enough time is available between two events for a single observer to be present at both, the time interval between the events that is recorded by a clock carried by that observer is called the *proper time*. There will then be a frame of reference in which this observer is at rest; in that frame both events will occur at the same location. That frame is called the *proper frame* for those two events.

In the example of the light pulse clock on the train, the train time, t_0, and the "factory" tick time, τ_0, are examples of proper times. The train frame, K_0, is a proper frame for the events consisting of the departure and the return of the light pulse, both of which occur at A_0 in that frame. A frames is a proper frame only in respect to a particular pair of events.

The time dilation equation, [7.8] is then an equation which relates the *proper time*, t_0, between a pair of events, measured in the *proper frame*, K_0, and the time, t, between the same pair of events measured in a frame K that is moving with respect to K_0 .

Eq. [7.8] makes it evident that t is always greater than t_o.

> The *proper time*, when it exists, is the ***shortest time*** interval between two events.

This means that, observed from any frame in which the events do *not* occur at the same place, the time interval is longer, or *dilated* compared with the proper time. The word, "dilated," means "stretched."

How to use ordinary clocks in a relativistic world

We are now prepared to return to the question of clocks.

A clock may be defined as a device that "ticks" in a regular way. Ticks are events, and the time interval between two consecutive ticks becomes a "unit time interval" for that clock.

The mechanism of the clock, whether it is a light pulse clock or a mechanical clock, determines the *proper time* of one "tick" of the clock. The difference between ordinary clocks and the light pulse clock is that the light pulse clock provides a way of calculating the duration of one tick (one round trip of the light pulse) in other frames of reference, whereas conventional clocks do not.

Having once derived the time dilation equation using a light pulse clock in a thought experiment, there is no longer any need for the light pulse clock.

Conventional clocks are *calibrated* in their at-rest frame, meaning that they have built into them a device which translates a count of "ticks" into a readout *based on the proper time of its tick*. This readout is therefore incorrect in all frames that have relative motion with respect to the clock. However, no matter what the mechanism of a clock, its tick time in a moving frame[3] can be calculated from its proper tick time using Eq 7.8.

If the clock on the train were an ordinary clock, the platform observer would reason that, since there is relative motion between her and the clock, she would observe *longer ticks* (ticks with more milliseconds) than the clock uses in calculating its readout. The factor γ, would correct the "factory" tick time of that clock by converting any readout time on the clock to the elapsed time in her frame of reference on the platform.

The saying that a "moving clock runs slower" can be confusing. All clocks that are similarly constructed run at the same rate. A clock observed at rest on the platform makes as many ticks per second as does a similar clock observed at rest on the train. A clock does not change the proper time between "ticks" by moving.

What changes when a clock is moving is the time observed between ticks by someone *not moving along with the clock*, or more appropriately stated, by an observer who is "moving with respect to the clock that is being observed."

Observer systems with synchronized clocks

A stationary observer can not, in fact, *read a moving clock*. To read a clock that moves from a point A to a point B on the platform (in the frame we call K), an observer would need to be at A when the clock is at A, and at B when the clock is at B.

For a single observer to do this, she would have to move with the clock, and doing that would put her in the frame of the moving clock, K_0. She would read the time interval, t_0, and not the time interval, t, in her own frame, K.

To read the times at A and at B from the point of view of frame K, without moving along with the clock from A to B, requires that

[3] that is, as observed in a frame that is in motion with respect to that clock

two observers be stationed, one at A and one at B (in frame K), with two clocks that had been previously synchronized in the platform frame. (See Chapter 6 Corollary 1c for methods of synchronizing two clocks in the same frame of reference.) This way two time readings would be recorded in frame K, and the difference between those readings would be the time interval, t. Such a measurement scheme is referred to as an "observer system."

The experiment with the light pulse clock obviated the need for an "observer system" on the platform, because the light path as seen from the platform measured the time interval directly. If platform clocks had been used, two would have been required.

If all motion is relative, is time dilation reciprocal?

I, standing on the platform, record more seconds between the emission (e_1) and the return (e_2) of the light pulse than you do, on the train. Which one of us is stationary and which is moving is arbitrary; the train and the platform move relative to each other, but either one (or neither) can be regarded as stationary.

Does this not imply that the comparison of seconds recorded should be reciprocal? That from your point of view, should you not record more seconds than I? This, of course, is impossible. These numbers can be written down and compared at leisure. If one is greater than the other, the other can not be greater than the first.

The answer to the question is, "No." The reason is that, while the ***motion*** is relative and reciprocal, there is not the same symmetry in the ***events***. The asymmetry in the events is not due to the fact that the events occur "on the train," and not "on the platform." The events occur at two places "in the universe," which can be observed from the platform or from the train or from a spy satellite high above the earth.

It is not where the events take place, but that the time dilation phenomenon was derived for the condition that in one frame of reference both events occur at the same location. If there is relative motion, it follows that they will not occur at the same location in the other frame. This condition does not flip around when we reverse the designation of which frame is stationary and which is in motion. Either way, the train frame is the one in which the events take place at the same location (where the clock is), and that is the frame in which the fewest seconds elapse.

Example 1: Pierre goes to Zurich

As the train pulls out of the Paris station, Pierre sets his watch by the station clock; it is 10 am. The train is due in Zurich at 4 pm. As his train pulls in at Zurich, Pierre checks his watch and finds that the train is very slightly early; it is several moments before 4 pm. Pierre is delighted to see Katrina already waiting at the platform. "It is wonderful that you came a bit before four o'clock," he says, "we made good time and arrived a bit early."

"Oh, but I did not come early. Your train came just as scheduled. See, it is exactly 4 o'clock," says Katrina, pointing to the Zurich station clock, which is synchronized with the Paris station clock. "Oh, no, that clock is wrong. It is not yet four o'clock," says Pierre, pointing to his watch.

Can you identify the two events of this problem?

Which clock(s) measured the proper time between these events?

Which clock(s) measured the dilated time between these events?

If the train's speed was 60 meters/sec (about 135 mi/hr) can you find γ and then the amount of time that the train was early?

(Ans: γ = 1.00000000000002 ; the train was 0.00000000043 sec early. Even at 135 mi/hr relativistic effects are insignificant! Later we will find examples in which relativistic effects are significant.)

Example 2: Timus Ybatu's Disputed Mile Run

Consider the historic mile run of Timus Ybatu[4], a few years back. Observers disagree whether Timus actually broke 3:50.00 or not. Even Timus conceded that he was considerably slower than the speed of light.

Fig 7.9 Timus Ybatu's controversial mile run

Timus and the officials had a long history of disagreement. Timus didn't trust them. He thought they had once already done him out of a world record. As a consequence, Timus had a stop watch made for him which was accurate to 10^{-14} seconds. Not to be outdone, the officials had a timing system of equal accuracy installed.

The officials installed two clocks of comparable precision (Fig 7.9), and synchronized them to correct for the predictable signal delay resulting from the fact that the start and finish of the mile run were one mile (1584.96 meters) apart, a distance which takes a light signal about 5.286724483 microseconds to traverse.

A light signal was sent from the start clock just as it was started at 0:00.000000000. When this signal arrived at the finish clock

---

[4] Timus Ybatu (Tim) is as fictional as is the watch he carries that keeps time to an accuracy of 10^{-14} seconds.

5.286724483 microseconds later, the finish clock was started with a reading of 0:00.000005286724483.

Timus carried *his* stop watch with him.

Timus ran his race. According to his stop watch, he ran the mile in 3:49.99999999999997. The officials recorded his time as 3:50.00000000000003. The difference in time was 6×10^{-14}sec. The ratio of the officials' time to Timus's was (1.000000000000000265).

They were, of course, both right. Timus's time was the *proper* time, t_0, since in the frame in which he measured the run, the beginning and the end point were at the same location. *He*, Timus, with his watch, was there at the beginning and at the end.

The officials recorded the time, t, in a different frame, K, not the proper frame. v, the relative velocity of the two frames was equal to Timus's average running speed of 1 mile (or 1600 meters) per about 230 seconds. (*Confirm, by using Eq. 7.8 that both times were indeed correct. Do not expect an ordinary calculator to deal with numbers to 18 digits; do calculations on the differences.*)

The world asked "which time is correct?"

How are we to decide what to put in the record books? (This question has to be decided by the world athletic committees.)

There is not one *correct* time. Not even the *proper* time is any more correct than any other. Proper time is unique, and is the shortest possible time measurement, and might therefore be a good time measurement to adopt as the official time, but it is not in any sense more correct than any other measured time interval.[5] You will have noticed that the official time that is recognized by Athletic Associations is not the proper time!

---

[5] There is a further matter involved here, that, in fact, resolves the issue rather nicely. We will take it up properly in a later section. This relates to the question of whether Timus, in his own frame of reference, did indeed run a whole mile. This is not easily answered with what we know so far. As far as the official record is concerned, the question that has an unambiguous answer is not, What was his time? but What was his speed? The judges have not realized this yet, however.

The story of Timus is frivolous, because of the absurdly small difference in the two measured times. It is, nevertheless, a perfectly accurate example of the effect of abandoning *absolute time* and living with Einstein's postulates.

Example 3: The muon that lived too long

The example involving Timus Ybatu's mile run is far-fetched, even though it is perfectly conceivable and correct. At the speed of even the fastest mile runner, the difference between the result from conventional, classical analysis and the result from a correct, relativistic analysis is trivial and negligible.

A runner who could move at somewhere near the speed of light would provide us a better example – one in which the relativistic solution is significantly different from the classical one. This happens (see Eq 7.8b) only if v/c is not so small as to be negligible; only then would the relativistic factor, γ, be significantly different from 1.

Of course there is no runner who moves that fast. But there is a "Timus Ybatu" of sorts who does, and who, like Timus, carries its own watch by which it measures the time of its run on a "track" at the ends of which *we* also keep two "fixed," synchronized clocks.

This new "Timus" is a little sub-atomic particle called a μ-particle, or, "muon." The "track" is the path from a point in the upper atmosphere about 15 km above the earth's surface, where a typical muon begins its life as the product of a collision between a cosmic ray and an atmospheric particle.

We will follow one such muon which happens to emerge from its birth process moving downward, toward earth's surface. The watch it carries is its "life clock." This watch ticks according to a process internal to the muon, analogous to the vibration of a quartz crystal – so, yes, it *is* a watch, which determines the likely time of the muon's demise. This "likely" time can vary; it is defined by a "half life," the time after which there is a 50% chance that it will have died - dissociated - and become something else.

Of course, the question arises whether its likely life span will give it enough time to reach the surface of the earth, where our laboratories can greet it, and count it.

A little biographical material on muons. They are, in some ways, like heavy electrons. They have one negative electronic charge, the same as an electron, and a mass about 200 times that of an electron — still far short of the mass of a proton or neutron.

Muons are highly unstable particles. As well as being made by collision in the upper atmosphere, muons can be artificially created by collisions in a laboratory. Once created they fall apart (*dissociate*) spontaneously with a half-life of about 2.5×10^{-6} seconds, or 2.5 μsec.

The collisions in the upper atmosphere usually produce muons with lots of kinctic energy; the typical upper atmosphere muon begins life with a speed of 99% of the speed of light, or $.99c$. This is about as fast as things can go; later we will show that one of the consequences of the relativity postulate is that nothing can go faster than the speed of light, c.

Typically a muon is produced at an altitude of 15km, or 15,000m above the surface of the earth.

At best, if it heads straight down toward earth, it would seem that the typical muon would have a pretty slim chance of making it down to our laboratories. You can figure the odds: travelling at a speed of 99% of 3×10^8 m/s, the time it would take the muon to reach us here at sea level is about $15000\text{m}/(2.97 \times 10^8\text{m/s})$ or about 50 μsec.

50 μsec is about 20 times the half-life of the muon. In each half-life, half of the remaining particles dissociate. After 20 half-lives only one in 2^{20} of the original number starting the trip from 15km up would still be a muon - the remainder will have dissociated into other things. One in 2^{20} is about one in a million.

As if by magic, however, the muons formed in the upper atmosphere appear to be invigorated, because their survival rate is more like one in eight than one in a million.

This is a very puzzling result. But it is puzzling only until you realize that in describing the problem as we have, we carelessly allowed the assumptions of absolute time and space to slip into our calculations. If we assert that "the half-life of the muon is 2.5 μsec," we have to ask, according to whose clock? We assumed that we could use our own clocks, according to which the muon would have to survive twenty half-lives to reach sea level in 50 μsec.

The muon's half-life, however, is determined by processes within the muon. These so-called "decay" processes are determined by a clock mechanism that is at rest in the muon's own frame of reference – like the watch carried by Timus Ybatu. The laboratory muons whose half life we measure to be 2.5 μsec, are slow muons (compared with the speed of light); the half-life of 2.5 μsec is therefore the *proper* half life of the muon.

Using the time dilation equation [7.8] we find that the muon's half-life, $t_{½}$, *in our frame*, (K), in which the earth is "fixed," is

$$t_{½} = 2.5\,\mu\text{sec} \times \frac{1}{\sqrt{1 - (0.99c)^2/c^2}} = 17.7\,\mu\text{sec} \qquad [7.9]$$

With a half-life (in the earth observer's frame) of 17.7 μsec, only about three half-lives are required for the muon to reach us from 15,000m up. In three half-lives, the survival rate is one in 2^3, or one in eight, which is just about what is observed.

This is a good occasion to re-examine the question, which time, or which half-life, is the "real" one? Which is the correct one?

Of course, while it is true that the "proper" half-life, the one measured in the frame of the particle itself, is probably the one that you would wish to list in a "Table of Half-lives of the Particles," it is not consistent to use it in a calculation that compares it with a survival-time requirement that is measured in the earth's rest frame. For that purpose one must use the half life measured in the earth frame, (the $t_{½}$ of Eq [7.9]).

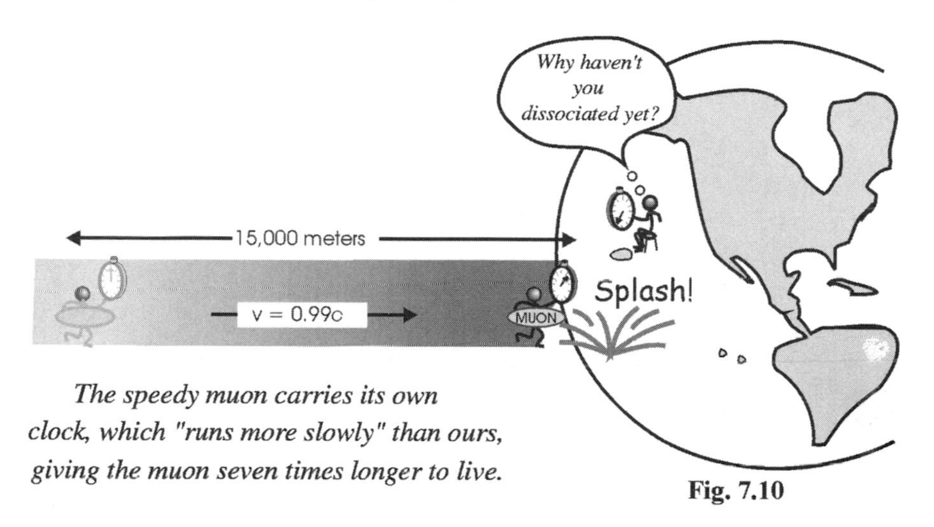

The speedy muon carries its own clock, which "runs more slowly" than ours, giving the muon seven times longer to live.

Fig. 7.10

For those who initially suspected that relativity is a product of optical illusion or experimental error, the muon experiment should make it abundantly clear that the prolonged half-life observed in the earth frame has no explanation in classical mechanics. These experiments were done with quite ordinary measurement devices, such as Geiger counters that measure particle populations at various altitudes. In no way can the difference between the laboratory and the in-flight half-lives of the muons be ascribed to instrument inaccuracy, or to artifacts due to signal delay or defective synchronization of clocks. These differences in time interval are consequences of the relativity postulates, and have no explanation in classical physics.

A problem remains: The muon itself does not have the benefit of the (dilated) half-life of 17.7 μsec that is observed from the earth's frame. In the muon's own frame its half-life is 2.5 μsec. How are we to explain to the muon that it has a one-in-eight chance of making it to the earth's surface? At the speed of 0.99c it can not travel 15,000 meters in three of its own half-lives. This mystery will be dealt with in the next chapter.

The Phenomena of Relativity

8.
Length and other kinds of Distance

The idea that distances depend on the frame of reference in which they are measured is not new. A moving and a non-moving observer see distances differently even in a world of fixed and immutable, non-relativistic space.

A person who walks up the aisle of a moving bus, may be seen to move a distance of 6 meters, or half a mile, depending on whether his motion is measured on the bus or along the street that the bus follows.

Fig 8.1 How far will the passenger move in the next 2 minutes?

Language and context are usually sufficient to sort out what distance is being discussed; we are rarely confused, and in any case the arithmetic is simple in these examples in classical mechanics.

Length

If we say that the person who walked the aisle of the bus from back to front "walked the length of the bus," the meaning will be generally understood, regardless of whether the bus was standing still or moving.

Length is a word that describes a particular distance. Length is a dimension of an object, although the "object" may be just a painted line, or even an imaginary line between two points.

Measuring the length of the bus, from outside the bus, would be easiest when the bus is standing still. A long tape measure could be stretched the length of the bus, and the two ends read. If the bus is moving, the task is made difficult by the motion of the bus, with both the front and the back of the bus moving past the tape measure. If your eye is quick and the bus not too fast, you could try to read both ends of the bus against the tape measure at the same time. If your eye is not quick enough, you might take a photograph with a high-speed camera. The camera is able to look at both points of measurement at once, in effect "freezing" the bus so that its motion does not matter.

A length measurement, in other words, requires the end points of the object to be observed *simultaneously*.

> *Length is a measure of the distance between two points observed simultaneously.*

Length under relativistic conditions

The issue of simultaneity is sharpened when the motion is rapid and relativity enters the picture. Because time is not absolute, but depends on the motion of the observer, simultaneity is frame-dependent. Two events that are simultaneous in one frame of reference, may not be simultaneous in another. (This is taken up in more detail in Chapter 14.)

Imagine a bus window that is 60 cm long when measured by a tape measure at rest with respect to the bus. That same window will not be 60 cm long if its length is measured holding that same meter stick up to the bus when it is in motion. This is a relativistic effect, and is not due to the technical difficulties of making such a very quick reading of the two ends of the window. The technical difficulties can give rise to illusions, which, however, can at least in principle be corrected. We will see that it is possible to arrange the practicalities of the measurement so that illusions are minimized, or eliminated. For example, if the geometry is arranged so that look-back times of both observations are the same, these can be made to cancel.

Measuring the length of an object that is at rest is easy. Since the alignment of the measuring device with the ends remains fixed, reading the ends slowly and not at once makes no difference.

If the object is in rapid motion, what strategy can insure simultaneity of the readings?

A moving observer makes simultaneous meter-stick readings

We will look at two possible schemes, not because anyone would ever actually carry them out, but because it may be helpful to see how one would, conceptually at least, implement the definition of "length" given above. Later, after we have developed the theory of length transformation, the same thing is usually achieved by calculation.

We will be measuring the *length of a stationary object*, a stripe painted on a "platform," using a *measuring device that is moving* by virtue of being on a moving "flatcar." (Note: The strategy would be similar if we were to measure the length of a *moving object* with a *stationary measuring device*.)

Suppose a flatcar moves past a platform on which is painted a stripe connecting two points, *A* and *B* (Fig 8.2). The platform is the frame in which the stripe is at rest. The platform is the *rest frame*, S_0, *for length measurements*. The distance *AB* measured in the rest frame is the *rest length*, ℓ_0 of the stripe.

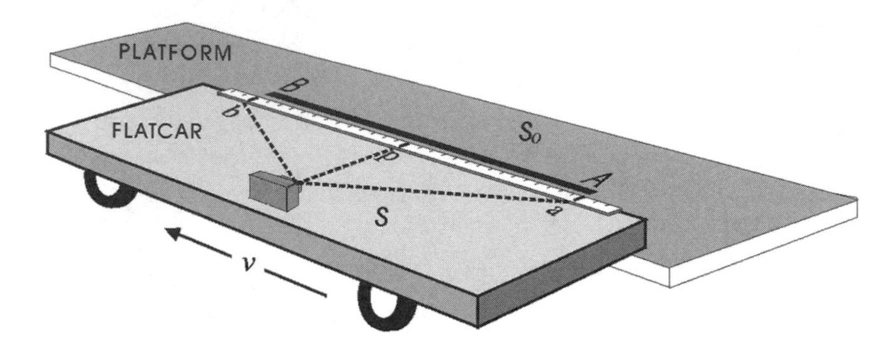

Fig 8.2 Scheme 1 for reading two meter-stick alignments simultaneously. Triangulation assures equal light-travel time from A and B.

To measure the *length, ℓ*, of the stripe *AB*, in the frame of the moving flatcar (*S*) one would read that length on a measuring stick fixed to the edge of the flatcar. One would have to read the two marks on the measuring stick that are aligned with the points *A* and *B* at the ends of the stripe on the platform *simultaneously*. Since the measurement of ℓ is being made on the flatcar, simultaneity has to be in the frame of the flatcar.

One could arrange an idealized high-speed wide-angle camera, placed as in Fig 8.2, to take photographs in rapid succession as the flatcar moves past the platform. Simultaneity of the alignment readings is assured by choosing that picture, out of the many taken, in which a point, *p*, on the measuring stick directly in front of the camera, is half way between the points *A* and *B* on the platform. In that one photograph the travel-time delay of the light from the two alignment points was the same, since the points are equidistant from the camera. Hence the light arriving at the camera left the two alignments simultaneously (in flatcar time). On the photo chosen in this way, one would read the two marks on the measuring stick attached to the flatcar that are aligned with points *A* and *B* on the platform. One would label these two marks *a* and *b*. The distance between *a* and *b* on the measuring stick in that photo is the length, ℓ, of the length *AB* as observed in the flatcar frame.

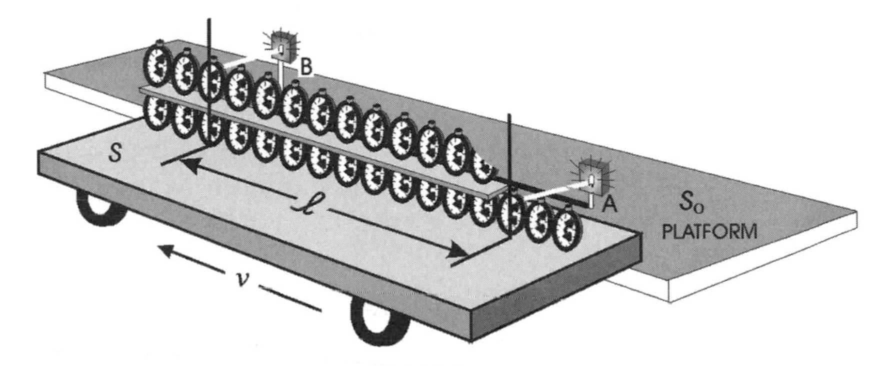

Fig 8.3 Alternative scheme. Two banks of synchronized clocks. ℓ is the distance between any two clocks that are stopped at the same time.

An alternative scheme (Fig 8.3) uses two banks of clocks fixed on the flatcar, all synchronized with each other, with one bank of clocks on top of the other, stretching the length of the flatcar. At *A* on the platform a light source emits a continuous beam towards the lower level of clocks; at *B* a continuous beam targets the upper bank of clocks. When a beam hits a receptor on a clock, that clock stops. This way the lower clocks are stopped as they pass *A*, and the clocks in the top bank are stopped as they pass *B*. Pairs of clocks stopped at the same time were at *A* and *B* simultaneously in the flatcar frame; the distance between such a pair of clocks, measured on the flatcar, is the *length* ℓ of the stripe, *AB*, *in the flatcar frame*.

The schemes of Figs 8.2 and 8.3 are ridiculously elaborate, and not very practical, but they suggest ways that one can conceptualize how these tricky simultaneous observations can be achieved. Such schemes define the meaning of "length" by specifying operations that can be performed to determine the value of such a length; such schemes are called "operational definitions."

"Length Contraction:" The relation between Rest Length and Length measured from a moving frame

The rest length of a stripe *AB* on a stationary platform is designated as, ℓ_o. The relation of that rest length to the length of the stripe, ℓ, as measured from a moving frame can now be determined by clocking the time it takes a mark **P** at the edge of a flatcar moving past the platform at a speed v, to move from its alignment with the point A to its alignment with the point B. A and B are points at the ends of the stripe (Fig 8.4).

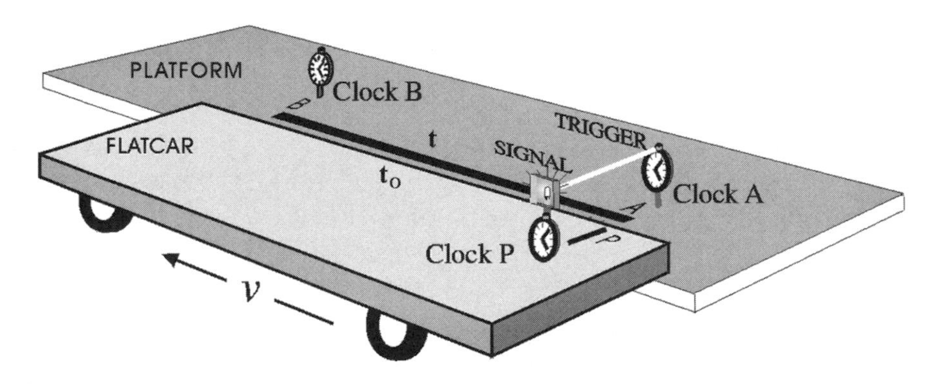

Fig 8.4 Scheme for finding length ℓ in the flatcar frame, based on the time it takes point P to go from alignment with *A* to alignment with *B*.

The basis of the scheme is corollary 4 (Chapter 6), which says that the speed, $v_{f,p}$, of the flatcar as seen from the platform is the same as the speed, $v_{p,f}$, of the platform in the view of an observer on the flatcar who considers the flatcar to be at rest.

$$v_{f,p} = v_{p,f} \qquad [8.1]$$

Clocks A and B at the end points of the stripe are synchronized in the platform frame (Corollary 1c). Each has a trigger which turns the clock off when a light beam from the clock at point P on the flatcar passes it. As the beam from clock P moves from A to B, it turns off clock B at a time t after turning off clock A. Since clocks A and B are at rest in the platform frame, the difference in clock readings, t, is the dilated time between the lining up of P with A and its lining up with B.

The proper time of travel, t_o, of point P from line-up with A to line-up with B can be measured on clock P by sighting these line-ups on the flatcar. Not easy to do, but no unusual scheme needs to be designed for that measurement.

These two times are related by the time dilation equation, 7.8.

$$t = \gamma t_o \quad [7.8]$$

The rest length, of the stripe can be measured at leisure in the platform frame.

The mutual and equal velocities $v_{f,p}$, and $v_{p,f}$, of the flatcar and the platform, are each equal to the ratio of distance to time in their respective frames of reference.

$$v_{f,p} = \ell_o / t \quad \text{and} \quad v_{p,f} = \ell / t_o \quad [8.2]$$

which, with [8.1] gives

$$\ell_o / t = \ell / t_o$$

With the time dilation equation, [7.8], we get

$$\ell_o / (\gamma t_o) = \ell / t_o$$

Multiplying both sides by t_o, gives the following relation between ℓ and ℓ_o, known as the "length contraction equation,"

$$\ell = \ell_o / \gamma$$

The length contraction equation

$$\ell = \ell_0 / \gamma \qquad [8.3]$$

where $\gamma = 1 / \sqrt{1 - v^2 / c^2}$

and ℓ_0 is the rest length

Eq [8.3] shows that the measured length of an object becomes shorter than the rest length by the factor γ when the object's length is measured by instruments in a frame that is moving with respect to the rest frame.

Eq. [8.3] shows that **the rest length, ℓ_0, is the *longest length* of an object**, since $\gamma \geq 1$.

What causes moving objects to be contracted?

It's a question that is nicely answered by C. Møller [*The Theory of Relativity*, Clarendon Press, Oxford, 1952, p46], "...the answer must be that such a question is just as delusive as if, after the discovery of the law of inertia, the question were put why a body left to itself will continue to move straight forward with uniform velocity."

The muon's journey, as seen by the muon

The mystery that remained at the end of the previous chapter was how the muon, in its own frame, where its half-life is 2.5 μsec, could be persuaded that it has a 1 in 8 (1 in 2^3) chance of reaching earth's surface from "15,000 meters up," travelling at 0.99c.

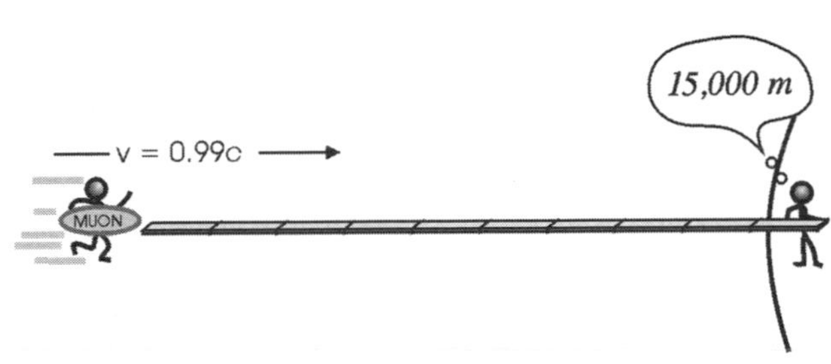

Fig 8.5 The measuring rod held fixed in the earth's frame reaches 15,000m to the muon.

For the earth observer, the mystery is solved by finding the dilated half-life, 17.7 μsec, of the muon in the earth frame. The muons are formed at an altitude of 15,000 m, as measured by earth observers, making that distance the rest length, ℓ_o.

At 0.99c it will take the muon 50 μsec, or three dilated half-lives of 17.7 μsec to reach the earth's surface.

To the muon, things look different. That 15,000 meter measuring rod that is fixed to the earth, as well as the earth itself, is contracted, due to the earth's motion with respect to the muon. The earth is not shrunk in all directions; it is contracted only along the direction of the relative motion of muon and earth. Earth becomes flattened, like a pancake.

The length of that earth-fixed measuring rod that reaches to the altitude where muons are made, is shortened according to Eq 8.3.

$$\ell = \ell_o / \gamma$$

$$= 15{,}000\text{m} \; / \; \{ \; 1 / \sqrt{1 - (.99c)^2 / c^2} \; \}$$

$$= 15{,}000\text{m} \times 0.141 \quad = \quad 2120\text{m}$$

The muon, at rest in its own frame of reference, (fig 8.6) sees the pancake earth approaching it from a distance of only 2120 m (not 15,000 m). Travelling at a speed of 0.99 c, pancake earth reaches the muon about 8.5 μsec later, about three times the half-life in the muon's own frame of reference. By the time the pancake reaches the muon, seven of eight of the muon's twins, born at the same time, have perished through natural radioactive dissociation, but one of the eight remains, in agreement with experiment.

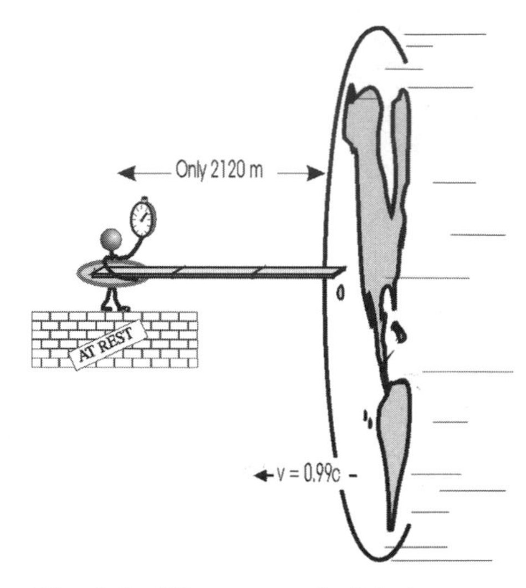

Fig 8.6 The muon holds its own measuring rod. Its half-life is short, but earth is 7 times closer.

It is of course necessary that the odds of a live encounter with earth must be the same in the view of an earth observer and in the view of a muon. The perseverance of muons to survive and have collisions with earth can not occur with greater likelihood when calculated in one frame than in the other – and it doesn't.

Length and Events – A Caution

It is tempting to think of the relation between contracted length and rest length as analogous to the relation between dilated time and proper time. There is an important difference.

Proper time is the time between two events occurring at one location. Dilated time is the time between the ***same two events*** measured in a frame in which the events have space separation.

One can think of two events as forming the space-time boundary of the interval between the events. Proper time and dilated time are the time components of the same space-time interval between the same pair of events. This is not true of the lengths of a line or object in two different frames. Length is, by definition, the space interval between two ***simultaneous*** events. If two events are simultaneous in one frame of reference, those same two events will not be simultaneous in another frame that has velocity with respect to the first.

And so, lengths in two different frames require the kind of analysis that we have shown above, and can not be obtained simply by applying relations that transform the space-time interval between the same two events from one frame to another.

The Lorentz transformation is precisely such a relation, and so it can not be used in general for length transformations. The length contraction equation [8.3] can be derived using the Lorentz transformation only with special reasoning. This derivation will be performed in Chapter 16.

Events in a journey

The events, (e_1) at the beginning and (e_2) at the end of a journey are the natural boundary events of a journey. Except in the journeys of massless particles, these two events are never simultaneous. The distance between these events is, therefore, not a ***length*** measurement. Length measurements are by definition simultaneous; the beginning and end of a journey are not.

The length of a line between New York and the Everglades in Florida might be of interest to a migrating bird that makes this journey annually, but this length would not be related to any space interval that might be computed from the events that constitute the beginning and end of the bird's journey.

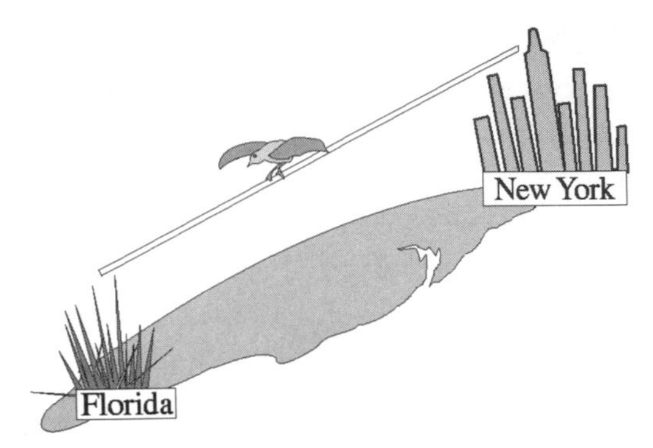

Fig 8.7 A bird holds a measuring stick to find the length of its flight from New York to the Everglades.

To determine the length of the trip as seen by the bird in flight would, according to the operational definition of "length," require the bird to carry in its claws a measuring stick thousands of miles long, and to sight the simultaneous alignment of the meter markings on this stick with New York and the Everglades. It would, of course, be easier for us to *calculate* this length by measuring the rest length and using Eq 8.3. (For the bird, this, of course, is also not practical!)

The length of this journey measured from the point of view of the bird is called [ℓ]. This length depends on the speed of the bird.

It is important to distinguish clearly between this length and the *distance,* [d], *between the beginning and the end of the journey, in the birds's frame.* Because the bird is present at both the beginning (e_1) and the end (e_2) of the journey, this distance is *always zero.*

In the traveller's frame, the distance between the beginning and the end of a journey is always zero

In general, for any traveller, be it a bird, a person, a particle, or any other object, [d] is zero, which is precisely what makes the traveller's own frame the proper frame for the journey, and the *time* e_1 to e_2 in the traveller's frame the proper time, t_o.

Because e_1 and e_2 are never simultaneous, the contracted *length* of the journey as seen by the traveller, is not related directly to the interval between the events. The journey's length, [ℓ], as seen in the frame of the traveller, is related to the rest length, [ℓ_o], through the length contraction equation, Eq 8.3.

[ℓ_o] is the rest length of the path of the journey, which is measured by laying down meter sticks that are at rest with respect to the end points of the journey. In the bird's journey, the meter sticks would be laid down on the ground over which the bird flies.

Because of the subtlety of the relations between the three distance quantities, [d], [ℓ_o], and [ℓ], we will recapitulate the muon example, and calculate these three quantities for two other examples, as well.

Example 1: The muon

[d] The muon begins its life in the upper atmosphere; let us designate that as event e_1. At some time later the muon has a collision with the surface of the earth; let us designate that as event e_2. The muon is present at both events; in the muon's frame, both events occur at the same place, and the distance, d, between the events, e_1 and e_2, and is ***zero***.

[ℓ_o] The rest length, ℓ_o, is the "tape measure" length (15,000 m) of that stripe in space that reaches from the location of the muon as it starts its journey, down to the earth's surface. The tape measure has to be held at rest relative to the earth

[ℓ] The contracted length, ℓ, 2120 m, is the muon's measure of the earth's distance from it at the time of e_1, when it is created "15,000 m" up. The muon must make a simultaneous reading of its own place and the place of the earth's surface on a long meter stick that it is holding - a tougher trick than even the bird of Fig 8.7 has to pull off!

To summarize: ℓ_o = 15,000 m; ℓ = 2120m; d = 0

Example 2: Flying from New York to San Francisco

Suppose that you travel by airplane at a speed of 1000 km/hr from New York to San Francisco. Maps indicate that *the distance* between these two cities is 5000 km. Let us find and distinguish the flight lengths, ℓ_0, ℓ, and the distance, d, between the start and end in the pilot's frame of reference.

[d] The space separation, or "distance" measured *in the pilot's frame* between your departure from New York and your arrival in San Francisco is zero. *In your own frame*, you have not moved (moved a distance of zero) while travelling from New York to San Francisco.

[ℓ_0] The rest length of a straight line between New York and San Francisco, ℓ_0, is measured in the frame, S_0, in which those cities are fixed. This is what is usually regarded as "the distance" between those cities and would be listed in an atlas or map. It is the distance that would be measured at leisure, using meter sticks (or kilometer sticks!) laid end to end, at rest, on the surface of the earth (ignore the curvature of the earth). This distance is 5000 km.

[ℓ] The *length* of the line between New York and San Francisco as seen from the airplane, is measured by taking simultaneous sightings from the moving airplane.

This length can not be measured with kilometer sticks leisurely placed end to end on the ground. In concept it would be measured by holding a 5000-kilometer long measuring rod in your hand while in the air above the continent and sighting (or photographing), simultaneously, the alignment of markings on this rod with each of the two cities, with the continent whizzing by below at 1000 km/hr.

What this measurement would turn out to be can be calculated from Eq [8.3],

$$\ell = \ell_0/\gamma = (5000\text{km}) / \{ 1 / \sqrt{1 - v^2/c^2} \}$$

With v = 1000km/hr (= 278 m/sec), ℓ is just .002 mm shorter than 5000 km.

To summarize: $\ell_0 = 5000$ km; $\ell = 4999.999999998$ km; $d = 0$

Example 3: Did Timus Ybatu really run a mile?

Recall that Timus Ybatu (Chapter 7), who carried his own stop watch, measured the time of his "mile run" to be somewhat less than what the judges declared to be his official time. Timus measured the time in his own frame, which gave him the "proper" time. The judges, using synchronized clocks in the earth ("platform") frame, obtained a time which was by the factor γ greater.

Timus was quite correct in asserting that his shorter time would have equal claim to be the "correct" time of his run. What Timus neglected to consider, and the judges, not being up on their physics, also failed to question, was whether Timus did indeed run a full mile in his own frame. The officials' mile was measured in the frame in which the starting and finish line are at rest, and is therefore the distance ℓ_0, in the S_0 frame, one mile exactly.

How far did Timus run in the frame in which he measured his own time? In *his own frame*, of course, Timus did not move at all between the beginning and end of his run. The distance, d, as always in these problems, is zero. But that is not the answer to the question. The desired distance is the ***length** measured in Timus's frame by **his** sighting of the start and the finish line, simultaneously.*

This length is ℓ, contracted from the rest length ℓ_0 that the judges measured. Timus's distance ℓ was one mile divided by γ.

Thus, while Timus ran the course by the factor γ quicker than the judges credited him, he also ran a course which was by the factor γ shorter. Thus by his own measurements in his own frame, he ran a course slightly less than a mile in slightly less time, with the result that he ran at the same *velocity* that the judges' figures credited him with. This should not be surprising, because the *relative velocity* between Timus and the track is the same, whether it is his velocity with respect to the track, or the track's velocity with respect to Timus (Corollary 4, Chap 6).

Timus would probably have been well advised not to press the issue. Though, by his own measurements, his time was shorter, his speed was *not* greater. In fact, he could have been disqualified on the grounds that according to those same measurements he did not complete a mile run, having run only 0.99999999999999735 mile.

To summarize: $\ell_0 = 1\text{mi}$; $\ell = 0.99999999999999735\text{mi}$; $d = 0$

Solution to Problem 5.1

Understanding how lengths vary as we view them from different frames of reference, we can now solve the problem that was posed at the close of Chapter 5.

Here is the problem: A 30 meter boat whose back end is 50 meters from a dock is travelling at ½c toward the dock. At this instant, a light pulse is emitted from the back end of the boat. A bomb at the front of the boat will be triggered if the light pulse strikes a light-sensitive triggering mechanism on the bomb. If the boat crashes into the dock before the trigger is set off, the bomb will be disabled by the crash, and will not explode.

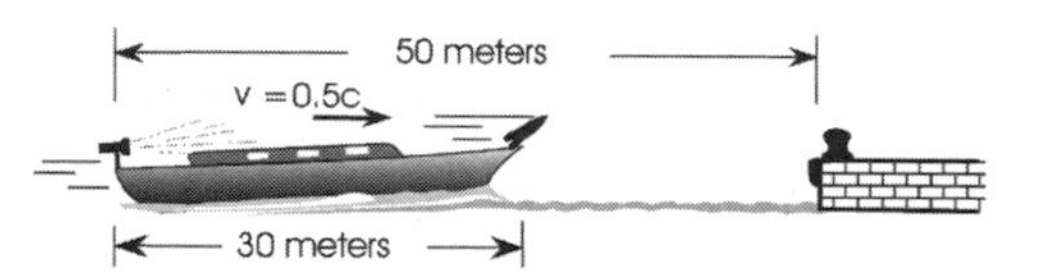

Fig 8.8 Problem 5.1. The question: Will the passengers survive?

The question is, will the bomb be set off, killing crew and passengers, or will everyone be safe?

By classical physics: two contradictory answers

In classical (Newtonian) mechanics, the answer depends on whether the light pulse travels toward the front of the boat at the speed of light *as seen from the boat* or *as seen from the shore.*

In classical physics, space and time are fixed and absolute. The boat is 30m long. It is (50 m – 30 m), or 20m from the dock. If the bomb does not destroy it first, the boat will crash into the dock after an interval of time equal to $(20\text{m})/(1.5 \times 10^8\text{m/s})$, or 1.33×10^{-7}sec.

(1) Classical physics solution: Light travels at "c" with respect to the boat.

If the light beam travels at a speed of *c* **with respect to the boat**, then the pulse must travel 30 m to reach the bomb. The time for the pulse to reach the bomb is $(30\text{m})/(3 \times 10^8\text{m/s})$, or 1.00×10^{-7}sec. This is less than 1.33×10^{-7}sec, the time it would take the boat to reach the dock The bomb will explode before the boat reaches the dock.

Classical Physics Answer #1: **ALL ARE KILLED!**

(2) Classical physics solution: Light travels at "c" with respect to the dock.

If the light beam travels at *c* **with respect to "fixed space"**, meaning in the frame of the shore and the dock, then it would take the pulse $(50\text{m})/(3 \times 10^8\text{m/s})$, or 1.67×10^{-7}sec to reach the dock. This is longer than 1.33×10^{-7}sec, the time it takes the boat to reach the dock. The boat will beat the pulse to the dock. The bomb will be knocked away and will not explode.

Classical Physics Answer #2: **ALL ARE SAFE!**

Only *one* of these two outcomes based on classical physics can actually occur. There is no way to decide which one, as long as we accept that light travels at the speed *c* in all frames, yet continue to solve problems on the assumption of absolute space and time.

This dilemma is what drove scientists to search for the "æther," the medium on which light travels at the speed *c*. If there were an æther it would determine which of the two classical solutions is correct, because it would determine in which frame the speed of light is c.

If the æther is at rest with the shore and dock, the speed of the light beam is $c - \mathrm{v}$ in the boat; everyone is safe. If the æther is dragged along with the boat, the speed of light is $c + \mathrm{v}$ with respect to the dock; the bomb would explode. If there were such a thing as æther, then here would be an experiment that could determine in what frame of reference the æther is at rest!

Solving it the right way gives the same answer in both frames

In the next two pages are solutions to the boat problem, using relativistic methods. The problem is solved twice. It is solved in the frame of reference of the dock, and it is solved again, this time in the frame of reference of the moving boat. While the numbers are quite different, because time and distance measurements are different in the two frames of reference, the answer to the question, "What happens to the bomb and to the passengers?" must be, and is, the same.

To solve the problem correctly, we have to get some clarification of the given dimensions. If a boat is said to be "30 meters long", it is reasonable to assume that this measurement was made while the boat was at rest, or by someone on the boat, whose frame of reference is the same as the boat's. It is also reasonable to suppose that the 50 meters distance was an observation made by someone on the shore, so that this measurement was made relative to the dock frame. The facts given in the problem were incomplete. Relativity helps us to ask for the information that was missing.

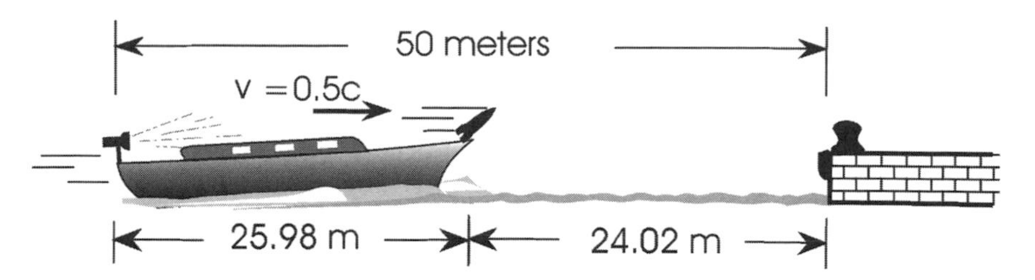

Fig 8.9 Relativistic Solution in the dock frame

(1) Relativistic solution in the frame of the dock

In a fixed frame, the distance from the rear of the boat to the shore is 50m. But the length of the boat is contracted.

$$\ell_{\text{BOAT}} = (30\text{m})/\gamma$$

$$\gamma = 1/\sqrt{1 - v^2/c^2} = 1/\sqrt{1 - (½)^2} = 1/\sqrt{0.75} = 1.1547$$

$$\ell_{\text{BOAT}} = (30\text{m})/1.1547 = 25.98\text{m}$$

The front of the boat is (50 m - 25.98 m) or 24.02m from the dock. Travelling at ½c, it will take the boat (24.02m)/(1.5×10^8m/s), or 1.601×10^{-7}sec to reach the dock.

It would take the pulse (50m)/($3\text{x}10^8$m/s) or $1.667\text{x}10^{-7}$sec to reach the dock. The pulse will not catch up with the front of the boat. The bomb will be knocked away when the boat crashes into the dock.

Relativistic Answer in the frame of the dock:
ALL ARE SAFE.

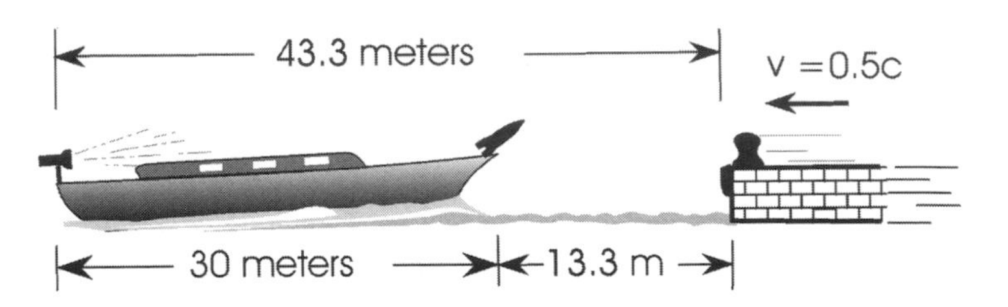

Fig 8.10 Relativistic solution in the boat frame

(2) Relativistic solution in the frame of the boat

In the frame of the boat, the boat is at rest. It is 30 m long. The space between the boat and the dock is contracted. The distance from the back end of the boat to the dock is $(50\text{m})/\gamma$.

The dock is moving at ½*c* toward the boat. The relative speed is the same as in (1), so γ is the same, or 1.1547. The dock is (50m)/(1.1547), or 43.30m from the back end of the boat.

But the front of the boat is 30 meters closer to the dock than the back of the boat. Hence, the distance the dock has to move (remember, the dock is moving; the boat is at rest!) is (43.30m - 30m), or 13.30m.

In the boat frame, it will take $(13.30\text{m})/(1.5 \times 10^8\text{m/s})$ or 0.887×10^{-7}sec, for the dock to crash into the front of the boat. (Note the time is not the same as it was in the fixed frame.)

In this frame, the pulse must travel 30m to reach the bomb; this would take $(30\text{m})/(3 \times 10^8\text{m/s})$, or 1.00×10^{-7}sec.

In 0.887×10^{-7}sec the bomb will be knocked away, *before* the pulse can reach it.

Relativistic Answer in the frame of the boat:
ALL ARE SAFE.

When solved correctly, the outcome, as it should, does not depend on who observes it.

The Phenomena of Relativity

9. Invariance of the Space–Time Interval

Is there *anything* at all that can be said without the qualifier, "as observed by ...?"

The word for something that doesn't change is, "invariant." We are looking for any characteristics of two events that do not depend on the point of view of the observer of those events. We are asking, are there any "frame invariant" quantities under the conditions of Einstein relativity?

In *classical* physics, the separation between two events is described by two quantities, each invariant, separately: the *time* between the events, Δt, and the *distance* between them, d. The displacement (in space) from one event to the other is a vector, $(\Delta x, \Delta y, \Delta z)$. Although the components, Δx, Δy, and Δz, depend on our choice of the placement and orientation of the x, y, and z axes, the distance, d, which is the magnitude of the displacement between the events, is independent of the placement of the space axes, and is given (according to Pythagorus' Theorem) by $d^2 = \Delta x^2 + \Delta y^2 + \Delta z^2$.

This distance between two points is independent of the way we choose coordinate axes to describe the location of the points. In fact, this distance exists quite in the absence of any coordinate axes at all. It is frame invariant because it is a *physically real quantity*, not just a marker on a set of axes that we happen to choose in describing the displacement.

The motivation in searching for frame invariant quantities is precisely this, that they are usually physical realities in a way that coordinates are not.

Degrees of freedom

In mathematical parlance, a relation between two variables is said to "reduce the number of degrees of freedom by one." In a system of simultaneous equations, a relation between two of the variables reduces by one the number of independent variables. For example, in a problem in which x, y, and z are independent variables, the relation z = 2y reduces the system from three to two independent variables. x remains independent, but if y is regarded as independent, then z is not, or vice versa.

In relativity, the speed of light is more than just the *speed* of something. The statement that this speed is a universal physical constant suggests that it is a relation between space and time. We might expect that this relation would reduce by one the degrees of freedom of the quantities defining the interval between two events. Instead of Δt and d, we should expect at most one physically real frame-invariant quantity. And since neither space nor time alone is such an invariant, we might expect that, if there is a frame invariant quantity, it would involve both time and space.

It is therefore an important question to ask whether there is, in four-dimensional space-time, a physical quantity associated with the interval between two *events* that exists whether or not we have a coordinate system to describe it, that does not depend on the methods and devices we use to give it a description?

Space-time "distance" between events

To develop the idea of a "space-time" interval between events, we look back at the use of the light pulse clock in chapter 7.

In a reproduction of Fig 7.7 below, are the views from the train and from the platform of two events, the departure and the return of the light pulse in the train. In the train frame, these two events both occur at the same place, at A_o. The distance between the events is zero; the time between the events is therefore the proper time, t_o.

In *any* other frame of reference (such as the platform frame) there is some distance d, and some time t, different from t_o, that separate these same two events. If the velocity of the train is v, then, from the point of view of the platform, the point, A_o, will have travelled a distance d from A to B, and this displacement will have taken time, t, in platform time. In platform variables, $v = d/t$. We know from the derivation in chapter 7, that $t = \gamma t_o$. Here, t is the dilated time, and γ has the usual meaning: $\gamma = 1/\sqrt{1 - v^2/c^2}$.

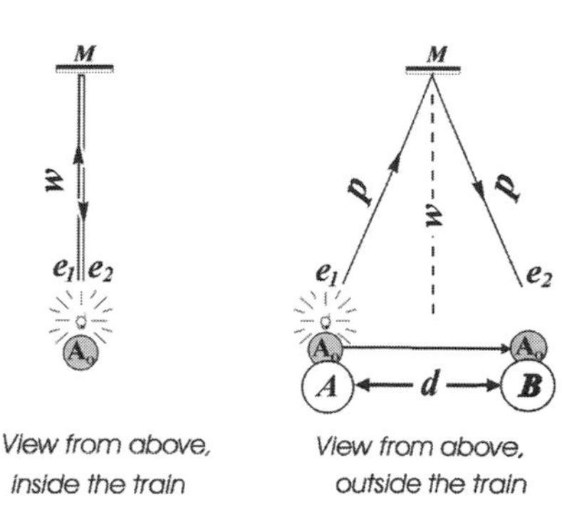

Fig 7.7 The path of the light pulse as seen by train and platform observers.

Rearranging the expression for γ,

$$1/\gamma^2 = 1 - v^2/c^2 = 1 - d^2/(c^2t^2)$$

Multiplying both sides by c^2t^2,

$$c^2t^2/\gamma^2 = c^2t^2 - d^2$$

Substituting $\gamma^2 t_o^2$ for t^2 in the left expression, and canceling γ^2,

$$c^2 t_o^2 = c^2t^2 - d^2$$

Finally, gives,

$$d^2 - c^2t^2 = -c^2 t_o^2 \qquad [9.1]$$

$c^2 t_o^2$ is a uniquely determined quantity because it involves a *particular* value of time, the proper time between the two events, t_o. What Eq 9.1 tells us is that, while the values of d and t may vary with the properties of the frame (the velocity, for example), the expression, $d^2 - c^2t^2$, always has the same value, because it is equal to the invariant quantity, $-c^2 t_o^2$.

The value of the expression on the left of Eq 9.1 is therefore *invariant* with respect to changes in frame of reference. ("All frames" in this context is to be taken to mean "all inertial frames.")

d^2 is equal to $\Delta x^2 + \Delta y^2 + \Delta z^2$. For convenience, the "Δ's" will be omitted, with the understanding that x, y, and z are intervals. Eq [9.1] is therefore,

$$x^2 + y^2 + z^2 - c^2t^2 = -c^2 t_o^2 \qquad [9.2]$$

The quantity, $-c^2 t_o^2$ is negative-definite. When two events have a proper time (that is when a single clock can move fast enough to be at both events), the positive quantity, $+c^2 t_o^2$, is given the name, τ^2.

τ^2 is referred to as (the square of) the "time-like interval" between the two events. It is (the square of) an *invariant space time interval*, and is written,

$$\tau^2 = c^2 t^2 - (x^2 + y^2 + z^2) \qquad [9.3a]$$

For any pair of events for which a proper frame exists, the expression on the right side of Eq 9.3a will be positive, τ^2 will be positive, and will have a real square root. Such a pair of events is said to have a "time-like" separation.

When τ^2 is negative

Not all pairs of events have a proper time. For some pairs of events, no clock can move fast enough to be at both events, so there is no frame of reference in which the events occur at the same location. In that case, there is no proper frame, and there is no proper time, t_o. For such a pair of events, the expression on the right side of Eq 9.3a is negative, and τ^2 has no real square root.

For such a pair of events there is, instead, a unique frame in which the *two events are simultaneous*. It would be natural to describe the interval between these two events as "space-like."

Let us designate as S_o the particular frame in which the two events are simultaneous, so that $t_o = 0$, and d_o is the rest length of a line that connects the locations of the two simultaneous events. The rest length is a uniquely determined quantity. Then it can be shown that in any other frame (unsubscripted)

$$c^2 t_o^2 - d_o^2 \quad = \quad c^2 t^2 - d^2 \qquad [9.4]$$

and since $t_o = 0$ (the events are simultaneous),

$$-d_o^2 \quad = \quad c^2 t^2 - d^2 \qquad [9.5]$$

We find that, as before, the quantity, $(c^2 t^2 - d^2)$ is an invariant with change of frame of reference, and could be given the same designation, τ^2. In the present case, where there is no proper time, τ^2 would be equal to $-d_o^2$, would be negative-definite and would have no real root.

The proof in this case is not as direct as the proof that established this quantity as invariant for events that have time-like separation. There is a proof, based on results we have not yet obtained.[1]

There is some expectation that the invariance of τ^2 extends to the domain ($\tau^2 < 0$), because it would be most unusual for a mathematical relation describing a physical fact to become untrue just because a root (the square root of τ) becomes imaginary[2]. The rule is likely to be applicable, but may require re-interpretation.

---

[1] A rigorous proof can be devised using the formalism of the Lorentz transformation (Chap 12).

[2] "imaginary" is used here in the mathematical sense, as a multiple of *i*, the square root of −1.

By taking the negative of both sides in [9.5], we write,

$$d_o^2 = d^2 - c^2t^2$$

$$\text{or} \quad d_o^2 = x^2 + y^2 + z^2 - c^2t^2 \qquad [9.5a]$$

In all frames of reference, the expression on the right of [9.5a] is therefore equal to d_o^2, a unique quantity, which is given the new name, σ^2 (Note that $\sigma^2 = -\tau^2$). Two events for which the expression on the right of [9.5a] is positive are said to be separated by a *space-like* invariant space-time interval, whose square, σ^2, is given by,

$$\sigma^2 = x^2 + y^2 + z^2 - c^2t^2 \qquad [9.3b]$$

Space-time interval between events

1. Time-like separation

There is a proper time, t_o.
There is a frame in which both events occur in sequence at the same location. A single clock can be present at both events.

$$\tau^2 = c^2t^2 - (x^2 + y^2 + z^2) \qquad [9.3a]$$

where $\tau^2 = c^2t_o^2$

2. Space-like separation

There is a frame in which the events are simultaneous, at different locations, separated by a distance that would be the rest length of a line connecting the locations.

$$\sigma^2 = x^2 + y^2 + z^2 - c^2t^2 \qquad [9.3b]$$

where $\sigma^2 = d_o^2$

With the exception of a unique borderline case (example 1 below) the separation between two events is either time-like or space-like.

What it means

The significance of these space-time intervals is that they are frame-invariant, *independent of who is making the observation*. The implication is that space-time is the place to look for absolute descriptions of events and the intervals between them. It suggests that space-time can be carved up into space and time in various ways; this carving up is akin to the various ways in which the x, y, and z axes in ordinary space can be shifted and rotated. The description of events in terms of space and time separately will depend on how this carving is done. Only when you leave space-time intact can you get in touch with the physical realities of the intervals between events. Events are physical realities, apart from how they are described; it is not surprising that the interval that separates them in space-time is also a physical reality, independent of the frame of reference in which it is measured.

A number of conclusions can be drawn from equations [9.3a] and [9.3b]. Some of these conclusions confirm observations we have already made. Others hold yet more surprises in store for us.

Example 1: The short (but eventful) life of a photon

On the boundary between time-like and space-like is the (seemingly) trivial case of two events that take place at the same time at the same place. There is a proper time that separates these two events, because the same clock can be at both events; the proper time is zero. There is also a space-like separation, because there is a frame of reference in which the events are simultaneous;

this space-like separation is also zero. For this special borderline case, $\tau^2 = \sigma^2 = 0$.

This example only *seems* trivial. It is not as trivial as it seems, because there are other frames of reference in which these same events are separated by both space and time.

$\tau^2 = \sigma^2 = 0$ means that,

$$x^2 + y^2 + z^2 - c^2t^2 = 0$$

$$x^2 + y^2 + z^2 = c^2t^2$$

$$d^2 = c^2t^2$$

$$d = ct$$

The relation, $d = ct$, describes an object travelling at the speed of light. It describes the motion of a "photon," a little burst of light wave, the elementary particle of light (or any other electromagnetic wave).

This example, then, describes the interval between the emission and the absorption of a photon. The *proper frame*, the frame fixed in the light particle itself, is then a frame in which the emission and absorption of the photon occur simultaneously, ($t_o = 0$) and at the same location ($d = 0$).

Imagine the life of a photon emitted by a distant star, that has travelled through the empty space of the universe for millions of years and has finally reached a photographic plate behind the lenses of one of our large telescopes. It is through such photons that we see stars. Because the photon has been on its way for millions of years, we see the star as it was millions of years ago, when that photon left the star.

In our view, the photon has travelled a distance, d, from its star to us in time t at a speed of c. Thus $d = ct$, and $d^2 = c^2t^2$, or,

$$x^2 + y^2 + z^2 = c^2t^2$$

Substitution of this in equation [9.3a] gives the result that $\tau^2 = 0$; substitution in equation [9.3b] gives the result that $\sigma^2 = 0$.

In earth's frame of reference the photon travelled distance *d* in time *t*. In the photon's own frame of reference, the location of its emission at the surface of the star, and the location of the photon's arrival on earth, are the same – the distance is zero! In the photon's own frame, the time of its emission and arrival are also the same – the time interval is zero!

From the photon's own point of view, it takes it no time at all to travel half way across the universe, or anywhere at all; and the distance that the journey covers is also zero. To the photon, the universe itself is the ultimately flattened super pancake, its size unaffected in the dimensions perpendicular to the direction of the photon's travel, but flattened to zero thickness in the direction of the photon's motion.

This has significant implications. One of these is the well-known unconditional stability of photons (as well as all other "massless" particles). The word "stability" is used here to mean it will not ever undergo *spontaneous dissociation*. (This does not mean it can not be absorbed by photographic film, for example, or by your skin or the surface of the earth.)

All particles that undergo spontaneous dissociation, do so with some characteristic half-life. You will recall that, as with the muon, the intrinsic half-life of a dissociating particle is always to be measured in its own frame of reference. In the case of the photon, no matter how short its half-life might be, it would be longer than the duration of the photon's existence, which, in its own frame of reference, is zero, no matter how far across the universe the photon may travel before interacting with something. The photon's stability is remarkable only from our point of view.

To the photon, its birth and demise are simultaneous. Its emission and absorption are simultaneous and contiguous events, and may as well be looked upon as a single event, in which the photon energy is transferred with no time delay from one object to another with which it is in contact.

The elusive question of neutrino mass

The unconditional stability of massless particles has come to figure prominently in recent attempts to determine whether the neutrino is a truly massless particle, or whether it has a very small mass. The great prevalence of neutrinos in the universe means that even if each neutrino has only a very small mass, the total mass of neutrinos in the universe could be sufficiently great to prevent the universe from expanding indefinitely.

Recent experiments have provided evidence that neutrinos may undergo particle transitions (called oscillations). Such transitions would occur with some intrinsic half-life. If this evidence is confirmed, it would rule out the possibility that neutrinos are massless; if they were massless they would be unconditionally stable.

Example 2: Proper Time

A conclusion that should not surprise us concerns events whose separation is time-like, for which the square of the space-time interval is,

$$\tau^2 = c^2 t^2 - (x^2 + y^2 + z^2) \qquad [9.3a]$$

which can be re-written,

$$t^2 = [\tau^2 + (x^2+y^2+z^2)] / c^2 = t_o^2 + (x^2+y^2+z^2) / c^2$$

From this equation one can conclude that the time separation between two events that are time-like is the least in the frame of reference in which the space separation $(x^2 + y^2 + z^2)$ is the shortest. For every pair of events that are time-like in their separation, there is a proper frame of reference in which the events occur in the same place. In that frame, the space part of the space-time interval, $x^2+y^2+z^2$, is zero, and therefore the time part $c^2 t_o^2$ of the interval, as well as t_o itself, is minimum. All other times are greater (dilated).

Example 3: Space-Time Interval Between the Birth and Crash of a Muon

Consider the life and death of the muon already discussed.

Let us suppose that we have a particular muon whose actual lifetime is three half-lives long. It thus has time enough to reach the surface of the earth from 15,000m up, or, in its own frame, to wait for the earth to fall down upon it from 2120m away.

The space-time interval between the birth and the crash of the muon, being frame-invariant, is the same whether viewed from the earth's frame (ours) or from the muon's frame.

The interval will be time-like. We can show that τ^2 is the same in both frames.

In the muon's frame, S, both events take place at the same location (the muon is always there)[3], so $d^2 = 0$, and

$$\tau^2 = c^2 t_o^2 \qquad [9.6]$$

where t_o is the proper time of the muon's flight.

In the earth's frame, S', we can represent the space part of τ^2 by $(d')^2$, where d' is the distance measured by earth observers between

---

[3] Note: in the muon's frame, the *space separation* between the events is zero, not the contracted *length*, d'/γ.

the upper atmosphere where the muon is born and the surface of the earth. This distance is 15,000 meters.

The distance d' is equal to vt'. v is the mutual velocity of muon and earth, and t' is the time as recorded by earth clocks between the birth of the muon and its crash into the earth's surface.

In the earth frame, the same (because it is frame-invariant) space-time interval is given by,

$$\tau^2 = c^2(t')^2 - (d')^2 \qquad [9.7]$$

Since $d' = vt'$, this equation can be re-written,

$$\tau^2 = c^2(t')^2 - v^2(t')^2 = (c^2 - v^2)(t')^2 \qquad [9.8]$$

The time dilation equation $(t')^2 = (t_o/\sqrt{1 - v^2/c^2})^2$, gives,

$$\tau^2 = (c^2 - v^2)(t')^2 = \frac{(c^2 - v^2)t_o^2}{(\sqrt{1 - v^2/c^2})^2} \qquad [9.9]$$

This simplifies to $\tau^2 = c^2 t_o^2$, which is the same as equation [9.6].

The Phenomena of Relativity

10. Relativistic Mass and the Speed Limit

It is not surprising that changing the basic assumptions about time and space would alter the perception of those two quantities, which have become two aspects of a combined quantity: space-time.

Perhaps it *is* surprising that mass, as well, is affected by the new perception of time and space. A second is no longer always a second; a meter is not always a meter. But a kilogram is also no longer a fixed property of a given object.

What are the observations by which we measure the mass of an object? There are two: (1) inertia, the reluctance of an object to change its velocity - its *sluggishness*; and (2) the response of an object to the presence of a gravitational field - its *weight*.

A moving object has greater inertia than an object at rest. Better stated: an object has greater inertia when viewed from a frame of reference that is in motion with respect to it than when it is viewed from the frame in which it is at rest.

It should be mentioned at the outset that not all physicists agree that the concept of a *relativistic mass* is necessary. Some maintain that Einstein never used the concept, but expressed all the necessary relations in terms of momentum and kinetic energy. These quantities he expressed in terms of the rest mass (denoted universally by the symbol m_o) and the parameters of the frame of reference in which they are observed.

It is obvious that the equivalence of mass and energy, which we will derive in the next chapter, makes one or the other of the two concepts, mass or energy, technically superfluous. It is common for particle physicists to express the mass of elementary particles in energy units, such as *electron Volts*. So, the use of *mass* in expressing the laws of physics is a convenience – to enable those who have grown up with the *mass* concept to write relations in familiar ways – but can be avoided altogether by the use of energy and momentum equations. Some scientists who hold the opinion that *relativistic mass* is a dubious concept maintain that the argument is more than just a disagreement over terminology[1].

In this book we choose to go along with the many other physicists who continue to use the concept of relativistic mass, for it has great utility in illuminating important and unquestionably true phenomena.

Relativistic mass

Relativistic mass is a concept that arises from relativistic *mechanics*, in contrast to most of the preceding concepts which are properly the domain of relativistic *kinematics*. Mechanics deals with laws of motion. The basic laws of motion are not derived from other principles; they are based on observation. In Newtonian mechanics the laws of motion are Newton's Laws. By a suitable definition of *momentum* it can be shown that Newton's Laws imply that momentum is conserved in isolated systems. Conservation principles for momentum and energy form an alternative set of fundamental laws of motion in classical physics; Newton's laws can be derived from the conservation principles for momentum and energy, just as the conservation principles for momentum and energy can be derived from Newton's Laws. It becomes a matter of preference which set of principles is regarded as more fundamental. They imply each other.

---

[1] See, for example, L.B. Okun, *Physics Today*, 42 #6, p31-36, June 1989.

In relativistic mechanics, the transformation equations for *Force* are complicated. It becomes preferable to frame the basic laws of motion in terms of conservation laws of momentum and energy.

Starting from scratch, it is not immediately obvious how momentum is to be defined in relativistic mechanics. It is desired that in the limit of slowly moving frames, the definition reduce to the classical definition, $\mathbf{p} = m\mathbf{v}$, where $\mathbf{p}$ is the symbol for momentum. Even though classical mechanics provides a starting point, there is no way to *derive* a relativistic definition of momentum. There is a certain reasonableness to the logic that follows, but it should be remembered that ultimately this logic does not constitute a derivation, but merely a path towards relativistic laws of motion that are ultimately justified, just as are Newton's Laws, by their faithful rendition of nature's behavior.

That this logic is not a simple or obvious extension of the postulates of relativity is evidenced by the fact that it was first proposed only in 1909, fully four years after Einstein published his Special Theory of Relativity[2].

The illustrative example which follows has been simplified and is therefore limited in its scope. More universal and more rigorous developments are more complicated without being, for the beginner, any more enlightening[3].

Beginning naively by defining momentum in the same way as in classical mechanics, $\mathbf{p} = m\mathbf{v}$, one is distressed to find that with this definition, the total momentum does not, in relativistic physics, remain constant in all isolated systems.

One has then the choice of retaining the familiar form and framing a new law in which momentum is *not* conserved *when it really ought to be conserved*, or of searching further in the hope that

---

[2] G.N. Lewis and R.C. Tolman, *Phil. Mag.* **18**, 510, 1909.

[3] See, for example, Arthur Beiser, *Concepts of Modern Physics*, 3rd Ed, McGraw-Hill, New York 1981; Chapter 1 section 1.7.; or C. Møller, *The Theory of Relativity*, Oxford Univ. Press, London, 1952; Chapter III Sec 26.

there might be a redefinition of momentum that obeys the momentum conservation law even under relativistic conditions.

The latter path is chosen, with the stipulation that the new definition of momentum must reduce to the classical definition, **p** = m**v**, in frames of reference in which **v** is (relativistically) small.

Suppose that on the edge of a moving flatcar of width w lies a ball, A, at rest (with respect to the flatcar). The flatcar frame is called S.

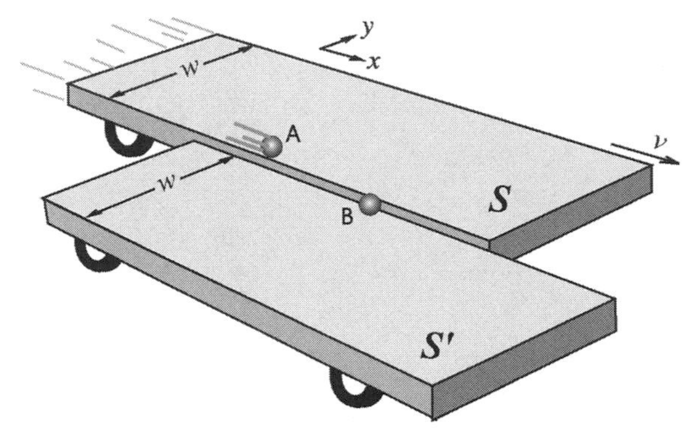

Fig 10.1 Flatcar S with ball 'A' is moving with velocity v. Flatcar S' is at rest.

As the train passes through a station at a great speed, v, it passes an identical train at rest on the adjoining track, in the "station" frame, S'. At the edge of a flatcar in the train that is standing in the station is an identical ball, B, at rest in the S' frame.

Designate the direction of the velocity of train S along the tracks as the x-direction, and the transverse direction, across the width dimension of both trains, as the y-direction.

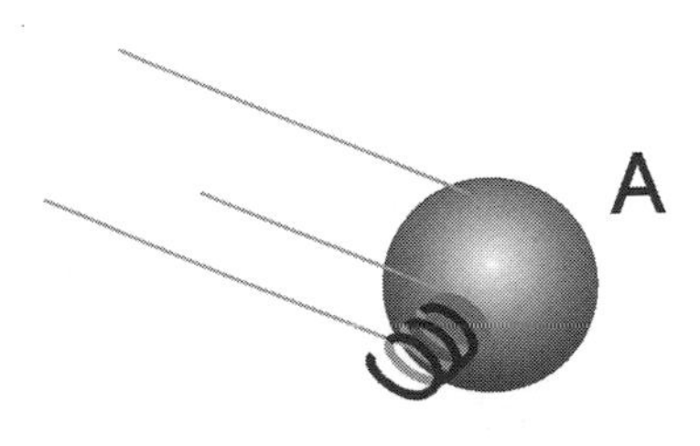

Fig 10.2 Just at the right moment, a fantasy spring pops out to push only in the y-direction.

As the S train passes the S' train, the balls come close enough to each other to collide.

To avoid the complication of momentum exchange in the x-direction, we create the fantasy of a small, massless, ideal spring attached to ball A. Just as the balls are opposite each other, the spring is released. It pops out toward ball B, pushing the balls away from each other, solely in the y-direction, across the width of their respective flatcars.

Neither ball has motion in the y-direction before the collision. If the momentum definition is to be classical in the limit of slow motion, the y-component of the total momentum of the two balls has to be zero before the collision. This has to be true in both the S and the S' frames. Write that as follows,

$$p_{A,y} + p_{B,y} = 0 \qquad [10.1]$$

$$p'_{A,y} + p'_{B,y} = 0$$

After the flatcars pass each other and the balls have the described collision, in which their x motion is unaffected, these relations must still be true if momentum is to be conserved. Again, this has to be true in both the S and the S' frames.

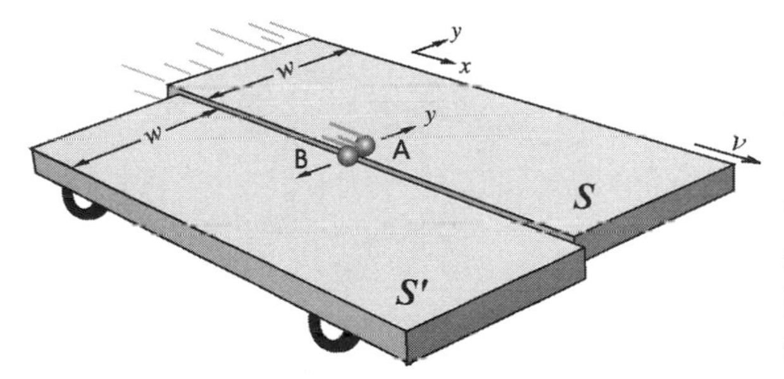

Fig 10.3 Because of the fantasy spring, the collision between ball A and ball B is entirely in the y-direction.

Suppose that ball A takes time, t_A (measured in S), to reach the far edge of flatcar S; then ball A's velocity in frame S is,

$$v_{A,y} = w/t_A$$

Ball B takes time t'_B (measured in S') to reach the far edge of flatcar S'; ball B's velocity in frame S' is,

$$v'_{B,y} = -w/t'_B$$

the minus sign indicating that ball B moves in the negative y-direction.

There is nothing special about the platform frame; S and S' are simply two frames with a mutual velocity of v. Because of the symmetry of the problem the same time is required for each ball to reach the far edge of its respective flat car. Provided that the time for each ball is observed in its own frame – the *A* ball in the *S* frame, and the *B* ball in the *S'* frame,

$$t_A = t'_B$$

The distances w in the y-direction are unaffected by the large velocities in the x-direction (Ch 6 Corollary 3). Consequently,

$$|v_{A,y}| = |v'_{B,y}| \quad [10.2]$$

A law of motion has to be written entirely in one frame of reference. The question then is not how $v_{A,y}$ and $v'_{B,y}$ are related, as shown in equation [10.2], but how $v_{A,y}$ and $v_{B,y}$ are related, or alternatively, how $v'_{A,y}$ and $v'_{B,y}$ are related.

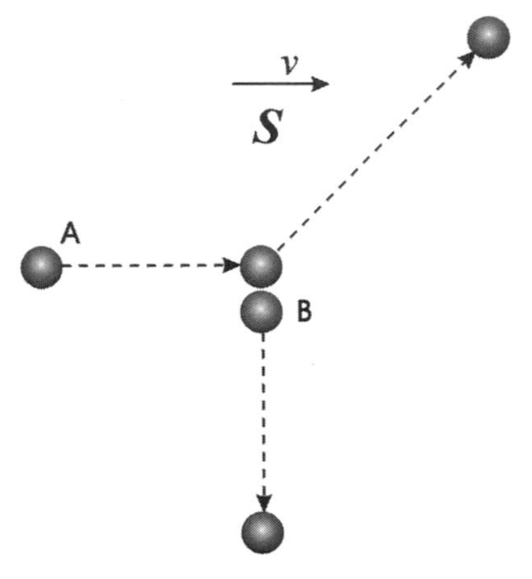

Fig 10.4 The collision seen from frame *S'*. Frame *S* is moving to the right at velocity v.

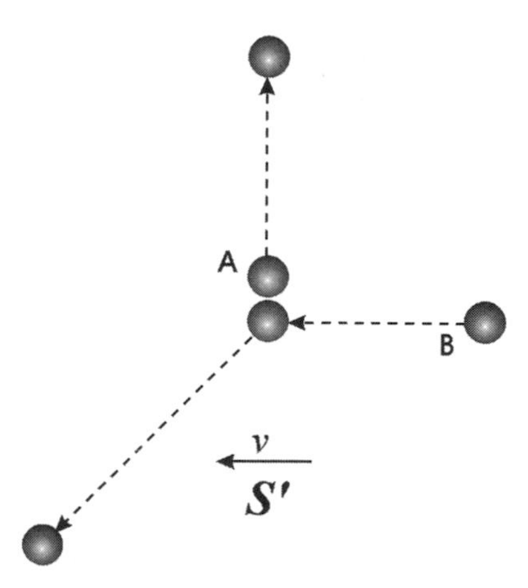

Fig 10.5 The collision seen from frame *S*. Frame *S'* is moving to the left at velocity v.

To answer these questions, let us place an observer in the *S'* frame, the frame of the flatcar with ball *B*, the frame that is at rest in the station. This observer sees the collision as pictured in Fig 10.4. The time taken by ball *A* to cross flatcar *S*, as seen from frame *S'*, is t'_A, the dilated time,

$$t'_A = \gamma t_A$$

where $\gamma = 1/\sqrt{1-v^2/c^2}$, and v is the velocity of the flatcar *S* with respect to frame *S'*.

The y component of the velocity of ball A as seen from the S' frame is therefore,

$$v'_{A,y} = \frac{w}{t'_A} = \frac{w}{\gamma t_A}$$

The y component of the velocity of ball B as seen from the same frame (S') is,

$$v'_{B,y} = -\frac{w}{t'_B}$$

But, we have already observed that, due to the symmetry of the problem, $t'_B = t_A$, so

$$|v'_{B,y}| = \frac{w}{t_A} = \gamma |v'_{A,y}| \qquad [10.3]$$

Part of the symmetry, as already stipulated, is that the two balls are identical. Suppose then that the mass of each ball, measured in its own rest frame (just in case that should make a difference!) is m_o. Then,

$$m_o |v'_{B,y}| = \gamma m_o |v'_{A,y}| \qquad [10.4]$$

The velocity, $v'_{B,y}$ that appears in the *left side* of equation 10.4 is the y component of the velocity of ball B observed in the frame of its own flatcar (S'). In this frame, the velocity of the ball has no x component. (All y components of velocity will be assumed so small that they do not contribute to the relativistically significant speed.) We required that the relativistic momentum reduce to the classical form when the velocities are measured in a frame in which these velocities are small. The left side of [10.4] is therefore the y component of momentum (absolute value) of ball B: $p'_{B,y}$.

The velocity, $v'_{A,y}$ that appears in the *right side* of equation 10.4 is the y component of the velocity of ball A, observed also in the frame of flatcar S'. For ball A this is the *other flatcar's frame*. This velocity is measured from a frame with a relativistically significant velocity (v) along the x direction with respect to the flatcar S in which ball A has little or no velocity. According to equation 10.4, then,

$$|p'_{B,y}| = \gamma m_o |v'_{A,y}| \qquad [10.4a]$$

According to equation 10.1, however, $|p'_{B,y}| = |p'_{A,y}|$, from which it follows that,

$$|p'_{A,y}| = \gamma m_o |v'_{A,y}| \qquad [10.4b]$$

It is important to notice that in [10.4b] the factor γ is related to the relative velocity, v, of the flatcars, in the x direction, and is not related to the (relativistically) small velocity, $v'_{A,y}$, of the ball A in the y direction. If momentum retains, in relativistic physics, the form of its classical definition, $\mathbf{p} = m\mathbf{v}$, we are led to conclude that the large velocity in the x direction has had the effect of increasing the mass of ball A from its value at rest, m_o, to a new value, (γm_o).

One speaks of m_o as the *rest mass* of an object, and designates as m' a quantity known as the *relativistic mass*, defined as follows,

$$m' = \gamma m_o = \frac{m_o}{\sqrt{1 - v^2/c^2}} \qquad [10.5]$$

Then it follows that the momentum is given by,

$$\mathbf{p} = \gamma m_o \mathbf{v}$$

or $\qquad [10.6]$

$$\mathbf{p} = m' \mathbf{v}$$

The *relativistic mass* increases without limit as v approaches c, making it ever more difficult to accelerate already rapidly moving objects. This result accounts, for example, for the decreased effect of electric and magnetic fields on rapidly moving electrons, as compared with the effect expected using non-relativistic equations.

Speed Limit

Equation [10.5] shows that, from the point of view of a fixed observer, an object's *inertia* increases as its *speed* increases. It becomes ever harder to increase its speed; finally, as the speed, v, approaches the speed of light (c), the denominator in [10.5] approaches zero, the inertia becomes very large, and the energy required to effect even small further increases in the speed becomes prohibitively large. In fact, as v approaches c, the energy required for further increase in speed goes up in such a way that infinite energy would be required to accelerate an object with finite rest mass to a velocity of c.

It is therefore said that an object whose rest mass is not zero can never achieve the speed c. Only particles with zero rest mass, such as photons, can reach this speed, and these particles travel *at exactly that speed*, a consequence of Maxwell's laws governing the propagation of electromagnetic fields.

Example 1

The muons of Chapters 7 and 8 were described as similar to electrons, but with about 200 times greater mass. Such descriptions are always in terms of rest mass.

$$m_{o,\mu} = 200\, m_{o,e} = 200 \times 9.11 \times 10^{-31}\text{kg} = 1.822 \times 10^{-28}\text{ kg}$$

The muons created in the upper atmosphere move at a speed of about $0.99\,c$. What is the relativistic mass of these muons?

Solution to Example 1:

$$\gamma = 1/\sqrt{1 - v^2/c^2} = 1/\sqrt{1 - 0.99^2} = 7.09$$

$$m_\mu' = \gamma \times m_{o,\mu} = 7.09 \times 1.822 \times 10^{-28}\ \text{kg} = 1.29 \times 10^{-27}\ \text{kg}$$

$$m_\mu' = 1418 \times m_{o,e}$$

As seen by those of us who are at rest, the mass of each of these muons is 1418 times (rather than 200 times) the rest mass of an electron.

The Phenomena of Relativity

11. Energy and its equivalence to Mass

*The subject of this chapter requires calculus. Readers not familiar with calculus can skip the first part, and begin reading with the section headed "**Summary of Energy results**" on page 148. From that point on no calculus is used in this chapter.*

Preliminaries: 1. Review of Momentum

In the previous section we *defined* momentum by two criteria: (1) in the limit of slowly moving objects it must approach the classical definition, and (2) it must be conserved in isolated systems. This led to the following:

$$\mathbf{p} = m'\mathbf{v}$$
$$\mathbf{p} = \gamma m_0 \mathbf{v} \qquad [11.1]$$

Preliminaries: 2. Force

The concept of force is not essential to a dynamics based on momentum and energy. But it is sometimes satisfying to the intuition to be able to picture motion as the result of pushes and pulls. Force can then be defined by extension of the Newtonian Force–momentum relation,

$$\mathbf{F} = \frac{d\mathbf{p}}{dt} \qquad [11.2a]$$

This equation reverts to Newton's Law in the form $\Sigma \mathbf{F} = m\,d\mathbf{v}/dt$ only in the classical limit. In general,

$$\frac{d\mathbf{p}}{dt} = \frac{d}{dt}(m'\mathbf{v}) = m'\,\frac{d\mathbf{v}}{dt} + \mathbf{v}\,\frac{dm'}{dt}$$

Relativistic Force does not have the absolute meaning that Force has in Newtonian mechanics.

That unique Force that is observed in the frame of reference in which the object is at rest (or moves slowly as a result of the Force) is called the Minkowski Force, $\mathbf{F}_M$. This force is the rate of change of the object's momentum with respect to the *proper time*, t_o, that is, the time in the frame in which the object is moving slowly,

$$\mathbf{F}_M = \frac{d\mathbf{p}}{dt_o} \qquad [11.2b]$$

Since $dt = \gamma\, dt_o$,

$$\mathbf{F} = \frac{\mathbf{F}_M}{\gamma}$$

Preliminaries: 3. Kinetic Energy

Kinetic Energy, K, can then be *defined* by formal extension of the Work-Energy Theorem, in terms of the Work (W), which is defined as $\int \mathbf{F}\cdot d\mathbf{r}$. Work produces change in Kinetic Energy, ΔK,

$$\Delta K = W = \int \mathbf{F}\cdot d\mathbf{r}$$

The Work-Energy Theorem can be written in terms of the rate at which Work is done,

$$\frac{d\mathbf{K}}{dt} = \frac{dW}{dt} = \mathbf{F}\cdot\frac{d\mathbf{r}}{dt} = \mathbf{F}\cdot\mathbf{v}$$

Substituting from [11.2a],

$$\frac{d\mathbf{K}}{dt} = \frac{d\mathbf{p}}{dt}\cdot\mathbf{v} = \frac{d}{dt}(m'\mathbf{v})\cdot\mathbf{v}$$

$$= \frac{d}{dt}(m_o\gamma\mathbf{v})\cdot\mathbf{v}$$

$$\frac{d\mathbf{K}}{dt} = m_o\gamma\left(\frac{d\mathbf{v}}{dt}\cdot\mathbf{v}\right) + m_o\frac{d\gamma}{dt}v^2 \qquad [11.3]$$

Since $\quad \dfrac{d\mathbf{v}}{dt}\cdot\mathbf{v} = v\dfrac{dv}{dt}$

and $\quad \dfrac{d\gamma}{dt} = \dfrac{d\gamma}{dv}\dfrac{dv}{dt}$

$$= \frac{dv}{dt}\frac{d}{dv}\left(1-\frac{v^2}{c^2}\right)^{-1/2}$$

equation [11.3] reduces to

$$\frac{dK}{dt} = m_o\,\gamma^3\,v\,\frac{dv}{dt}$$

which finally becomes

$$\frac{dK}{dt} = \frac{d}{dt}(\gamma m_o c^2)$$

Integrated, this becomes,

$$K = \gamma m_o c^2 + \textit{constant}$$

Setting $K=0$ when $v=0$ ($\gamma=1$), it follows that the *constant* in this equation is equal to $(-m_o c^2)$, and therefore,

$$K = m' c^2 - m_o c^2 \qquad [11.4]$$

It is useful to turn this equation around, and write it this way,

$$m' c^2 = m_o c^2 + K \qquad [11.4a]$$

In this equation, $m_o c^2$ is an object's "rest energy," E_o, the energy it has in the absence of motion energy. When its motion energy, or Kinetic Energy, is added to that rest energy, the result is $m' c^2$, which is its total energy, E.

Summary of Energy results

We read Eq 11.4a as follows: The total energy E is the sum of the energy that the object has when it is at rest, E_o, and the energy the object has as a result of its motion, K.

$$E = E_o + K \qquad [11.5]$$

The rest energy is

$$E_o = m_o c^2 \qquad [11.6]$$

The total energy is

$$E = m' c^2 = \gamma m_o c^2 \qquad [11.7]$$

Consequently the Kinetic Energy is

$$K = E - E_o = (\gamma - 1) m_o c^2$$

$$\text{or} \quad K = \left[\frac{1}{\sqrt{1 - v^2/c^2}} - 1 \right] m_o c^2 \qquad [11.8]$$

It would be awkward indeed if this expression for the kinetic energy did not, for small velocities, reduce to the classical expression. In the limit of $v \ll c$, the square root in the denominator of γ can be approximated, giving,

$$\frac{1}{\sqrt{1 - v^2/c^2}} \approx \frac{1}{1 - \frac{1}{2} v^2/c^2}$$

$$\approx 1 + \tfrac{1}{2} v^2/c^2 \qquad [11.9]$$

Applying the approximation [11.9] to [11.8], one obtains,

$$K\ (@v \ll c) \approx \{ 1 + \tfrac{1}{2} v^2/c^2 - 1 \}\ m_o c^2$$

$$\approx \tfrac{1}{2} m_o v^2 \qquad [11.10]$$

Thus, the total energy of an object that moves slowly (with respect to the observer) is the sum of the rest energy, $m_o c^2$, and the Kinetic Energy, calculated from its usual form, $\frac{1}{2} m_o v^2$. At great speeds, the total energy is $\gamma m_0 c^2$, and the Kinetic Energy must be calculated from equation [11.4].

$E = mc^2$: the equivalence of mass and energy

It is evident from these relations that there is an intimate connection between mass and energy. For every Joule of energy there is a certain fraction of a kilogram of mass; likewise every kilogram of mass entails some number of Joules of energy.

It is from these results, equations [11.6] and [11.7] in particular, that the equivalence of mass and energy, and the famous equation, $E = mc^2$, derives. The inertia, or sluggishness, due to mass, as well as the gravitational property of mass, are in this view simply manifestations of the presence of energy.

The equation, $E = mc^2$, has the undeserved reputation of *causing* the release of energy in nuclear bombs and in nuclear reactors, or of *explaining* it. In fact, the role of this equation in the development of nuclear technology was much more modest. $E = mc^2$ provided the means by which the enormous energy available from nuclear reactions could be *measured*, that is, by weighing it.

The energy released in nuclear reactions is so large that the mass of that energy is large enough in comparison with the masses of the nuclei involved that the weight of that energy can be determined by actual weight measurement on a laboratory balance. The weight of the energy can not be measured by itself, but it can be calculated from the difference of the weight of the nuclei before and the weight of the nuclei after the nuclear reaction. Long before relativity provided an explanation for it, this "mass defect" had been known to chemists as the troublesome difference in mass of the same total number of protons and neutrons arranged differently in the nuclei of different elements.

Relativity explained the "mass defect" as the mass of the binding energy released during the nuclear rearrangement. It is the energy difference between the nuclear binding energy (which is a potential energy of nuclear attraction forces) of the reactant nuclei and that of the product nuclei. During a nuclear reaction the "packaging" of the neutrons and protons changes. If the packing becomes tighter, or if new bonds are formed, energy is released.

This in itself is not remarkable. When oxygen and carbon combine in the burning of coal, the oxygen atoms form chemical bonds with carbon atoms, and it is the formation of these bonds that releases some of the chemical bond potential energy in the form of heat. The same thing happens in nuclear reactions, except that the quantities of energy released are vastly greater.

For example, in a uranium fission reactor, when a nucleus of Uranium-235 is hit with a neutron, it can split into two smaller nuclei, such as Barium and Krypton. Because the Uranium nucleus is large, and the nucleons at opposite sides of the nucleus are relatively far apart, the attractions of the nucleons for each other is barely able to hold that nucleus together.

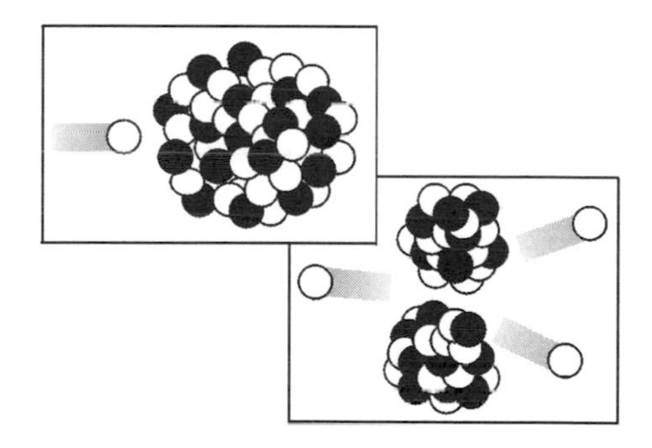

Fig 11.1 Uranium-235 splits into two smaller nuclei, in which the "packing" of nucleons is tighter, releasing bonding energy.

This is why Uranium-235 is fissionable. It also means that when those same nucleons are repacked in the smaller fragment nuclei, they pack more tightly, releasing energy. Most of this energy is released in the form of kinetic energy of three very fast neutrons. This kinetic energy can be absorbed by water in the reactor heat transfer system, eventually boiling into the steam that drives the turbines which in turn power the electric generators.

In the hydrogen fusion reaction, which occurs in the sun and in thermonuclear bombs, a nucleus of deuterium (hydrogen with an extra neutron) is caused to "fuse" (merge) with a nucleus of tritium (hydrogen with two extra neutrons), forming a new bond and making helium. The formation of the new bond releases an enormous amount of energy, again usually in the form of kinetic energy.

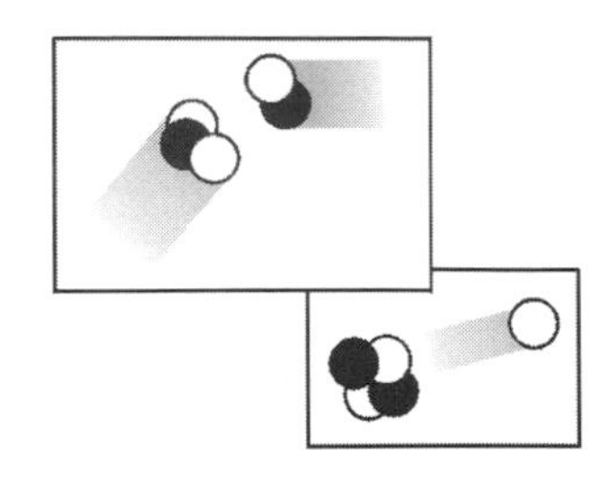

Fig 11.2 Tritium nucleus (top, left) welcomes a second proton, bonds tightly with it.

The role of $E=mc^2$ in nuclear technology

The isotope tables that chemists had accumulated through ever more careful measurements of the weights (and masses) of the elements and their isotopes gave nuclear physicists the total masses, before and after, for the reactants and products in possible nuclear reactions, and the energy difference between these, the so-called mass defect for the reaction. The understanding that the mass defect represented the energy released was extremely important, because it enabled early nuclear physicists to calculate the energy available from a particular nuclear reaction, if it could be made to go. For obvious reasons, it was not practical to obtain this information in the way one finds the energy released per gram of gasoline burned, by making the reaction go and measuring the heat produced.

Mass is not "converted" to energy in nuclear reactions. The mass that is involved *is* energy both before and after. Before the reaction, the mass that becomes the "mass defect" is the mass of the nuclear bond potential energy. During the reaction, as new bonds are formed, or existing bonds are tightened or replaced by stronger bonds, this energy is most commonly released as kinetic energy, in the form of heat and/or mechanical energy.

In its most general sense, the equivalence of mass and energy states that all energy, be it kinetic energy or potential energy, exhibits both the gravitational and inertial properties of mass. This idea is difficult to grasp; it seemed odd even to Einstein when he first found that it follows logically from the premises of his theory of relativity. He confided to a friend at the time that, "...this thought is both amusing and attractive, but whether the Lord laughs at me concerning this notion and has led me around by the nose – that I cannot know."[1]

Example 1: A fast ball gains mass

A ½kg baseball is pitched at a speed of 40m/s (about 100mi/hr). Its kinetic energy, K, contributes additional inertia and weight to the ball, amounting to the mass equivalent of the kinetic energy, K.

The increase in mass is therefore K/c^2, or $(\frac{1}{2}mv^2)/c^2$ (at small velocities, the classical expression for kinetic energy is very nearly correct). The increase in mass is $\frac{1}{2}(\frac{1}{2}\text{kg})(40\text{m/s})^2/(3\text{x}10^8\text{m/s})^2$, or 0.0000000000000444kg, hardly an amount to worry over.

An electron travelling at half the speed of light, however, has a mass increase due to its kinetic energy that is not a negligible fraction of its original, or "rest" mass.

Example 2: Energy to increase speed of an electron by 1 m/s

How much *Work* is needed to increase the speed of an electron that is initially moving at $0.9c$, by 1m/s? The rest mass of an

---

[1] letter to Conrad Habicht, widely quoted, eg, in Hey and Walters, "Einstein's Mirror," Cambridge University Press, Cambridge, UK, 1997.

electron is 9.11×10^{-31}kg. The Work needed is the difference between the kinetic energy at $v = 0.9c$ and the kinetic energy at $v = (0.9c + 1\text{m/s})$.

This example demonstrates how substantial a difference relativity can make when the speed is great. This will be shown by solving the problem (inappropriately) using classical mechanics, and then solving the problem (appropriately) using relativistic methods.

Method 1: classical mechanics (incorrect)

$$\begin{aligned}\Delta K &= \tfrac{1}{2}m(.9c + 1\text{m/s})^2 - \tfrac{1}{2}m(.9c)^2 \\ &\approx \tfrac{1}{2}(9.11 \times 10^{-31}\text{kg}) \{ (2)(.9)(3 \times 10^8\text{m/s})(1\text{m/s}) \} \\ &= 2.46 \times 10^{-22} \text{ Joules}\end{aligned}$$

Method 2: relativistic mechanics

The speed of the electron is great ($0.9c$), and so the classical solution is not a good approximation. To solve this problem *correctly*, use equation [11.8].

$$\Delta K = \left[\frac{1}{\sqrt{1-(v')^2/c^2}} - 1 \right] m_o c^2 - \left[\frac{1}{\sqrt{1-v^2/c^2}} - 1 \right] m_o c^2$$

$$\Delta K = \left[\frac{1}{\sqrt{1-(v')^2/c^2}} - \frac{1}{\sqrt{1-v^2/c^2}} \right] m_o c^2$$

where $v = 0.9c$ and $v' = v + 1\text{m/s}$. The result is

$$\Delta K = 2.97 \times 10^{-21} \text{ Joules}$$

which is about 12 times as great as the value calculated above from classical mechanics.

12. The Lorentz Transformation

In the chapter after this one, we intend to take up a question that has been hanging over us ever since, in Chapter 4, we learned how Einstein salvaged both the principle of relativity (that laws of physics take the same form in all inertial frames of reference) and "the simple law of the propagation of light *in vacuo*," (Maxwell's Laws). To hold on to both of these seemingly contradictory principles, Einstein sacrificed the classical theorem of velocity addition, Eq 3.2, and with that also went the assumption of absolute time and fixed space.

The question is, what replaces that classical velocity addition equation under the conditions of Einstein's relativity postulates?

Finding an answer to that question, we would come very close to deriving a more general relation, known as the Lorentz Transformation. Our plan, then, is to perform that more general derivation in this chapter. A small detour from that derivation, in the next chapter, will then easily give us the relativistic velocity addition rule that we are looking for. Later, when we understand better the nature of transformations, we will return to applications of the Lorentz Transformation.

The (relativistic) Velocity Addition Problem

Let us begin by restating the velocity addition problem. Using the image of the train problem of Chapter 7, consider an object moving with a velocity, v, in a "train" frame, K. The "train" frame is moving with velocity, u, with respect to a fixed, or "platform" frame. The question is, what is the velocity, V, that a fixed observer would ascribe to the object in the train?

According to the classical theorem of velocity addition, Eq 3.2, the results would be that V is the simple sum of u, the "train" velocity, and v the velocity of the object with respect to the train:

$$V = u + v \qquad [3.2]$$

For fast moving objects, this is not correct.

To recast the problem with imagery of faster-moving objects, let us set the problem in the rocket pictured in Fig 12.1. The rocket is moving with respect to a "fixed" (earth) observer with velocity, u. Inside the rocket, a bullet has been fired that is moving with speed, v, with respect to coordinates fixed in the rocket. From the point of view of the fixed observer outside the rocket, how fast is the bullet moving? If, for example, the rocket's velocity with respect to the fixed observer were ¾c, and the bullet's velocity with respect to the rocket were also ¾c, the classical velocity addition rule would predict that, seen from a fixed frame, the bullet's velocity would be 1.5c – exceeding the speed of light, an impossibility. Obviously, relativity must provide us with a different velocity addition theorem.

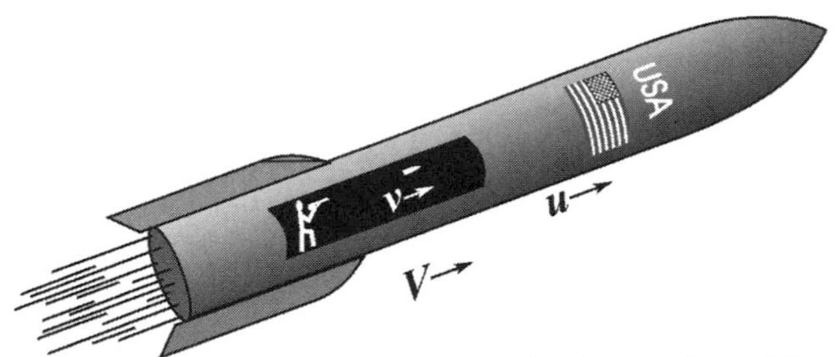

Fig 12.1 A bullet is fired inside the rocket. The bullet speed is v. The rocket has velocity, u. How do these velocities add?

If we translate the velocities, u and v, that are the givens of this problem into displacements over some interval of time, the problem will take on the appearance of the time dilation problem of Chapter 7. This will suggest that the method of that derivation may have some hints for us in our present problem.

Inside the rocket

Inside the rocket, one observes the bullet in two locations. The bullet at a is called event e_1. Event e_2, occurs at a time, t, later, at point, b. If the distance between a and b in the rocket is called, x, then, the relation between these variables and the bullet's velocity, all in the rocket frame, is,

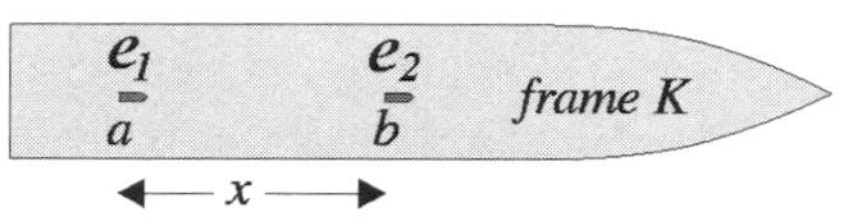

Fig 12.2 It takes the bullet time, t, to go from a to b in the rocket. t is measured in the rocket frame.

$$x = v\,t \qquad [12.1]$$

Outside the rocket

The events, e_1 and e_2, of the bullet's motion from a to b inside the rocket can also be observed from outside, from the "fixed" frame, K'. The time and distance intervals, t' and x', between these two events, will be different in K'.

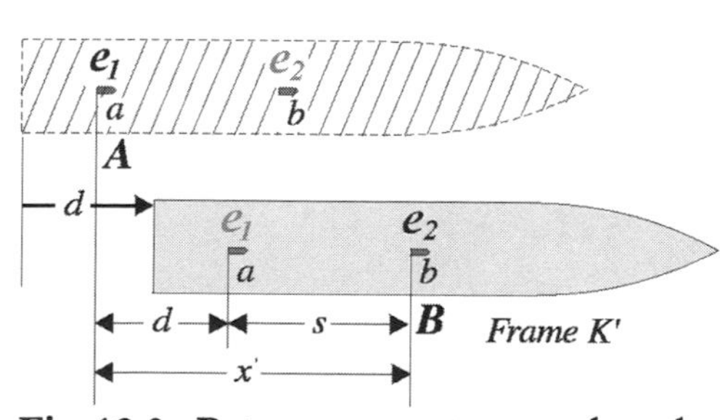

Fig 12.3 Between events e_1 and e_2 the rocket has moved a distance d, in the fixed frame, K'.

In the time, t' (unknown at this point), between the events, e_1 and e_2, the rocket will have moved forward a distance, d, in the fixed frame. This distance is determined by the rocket speed, u.

$$d = u\,t' \qquad [12.2]$$

The distance that the bullet has travelled, as seen from outside the rocket, in the fixed frame, is x' (unknown at this point). This is the distance between two "paint spots" that might, in fantasy, have been left by the bullet on the "platform" of outer space. (You may recall the paint spots on the platform that were used in the logic of Chapter 7.) One "paint spot" is at A (Fig 12.3), the location in the fixed frame opposite the bullet at the time of event e_1, when the bullet is at a; B is the location of a "paint spot" opposite the bullet when it is at b, at the time of event e_2.

The distance, x', is important to us, because it is related to the velocity, V, of the bullet as seen from outside the rocket, in K'.

$$x' = V t' \qquad [12.3]$$

For future reference, from Fig 12.3 the distance, x' is also the sum of d and s.

$$x' = d + s$$

s is the contracted value of the rest length of the line ab in the rocket. The length $|ab|$ in the rocket is equal to the distance, x, that the bullet travelled in the rocket. From the length contraction equation [8.3],

$$s = |ab| / \gamma = x / \gamma$$

Then, with [12.2],

$$x' = u t' + x / \gamma$$

$$\text{where} \quad \gamma = 1 / \sqrt{1 - u^2 / c^2} \qquad [12.4]$$

The Strategy

The relativistic velocity addition problem is: Given u and v, find V.

Eq 12.1, $x = v\,t$, allows us to re-state the problem so that u, x, and t are given, instead of u, v, and t.

The problem now is: **given u, x, and t, find t' and x'**. The solution to this problem is the **Lorentz transformation.**

Once we have found t' and x', we can easily return to the original problem, using Eq 12.3 to find V.

In the re-stated problem, the two events, e_1 and e_2, are to be looked at from two points of view: in the rocket frame, K, and in the fixed frame outside the rocket, K'.

We want to find how the time and space intervals between these events transform as we go to the fixed frame, K'. The task is to find expressions for x' and t' in terms of x, t, and u.

> ***In classical physics,*** we would assume that $t' = t$. Then, by the geometry of Fig 12.3, x' would be $d + x$, where d, the distance travelled by the rocket in time t' would be ut. This would give, $x' = x + ut$.
>
> ***But that's in classical physics!*** Not true if speed is great.

To solve the problem relativistically, we proceed as we did in Chapter 7, where we derived the time dilation equation by using the device of a light pulse clock. In the place of the classical assumption, $t' = t$, we impose the condition that the speed of light is the same in all frames of reference. This is accomplished by calculating time by the path length of a light pulse.

We can use a light pulse clock of essentially the same design that worked so well for us in Chapter 7. However, the use of this device will be complicated by the fact that in the present problem neither the rocket frame nor the fixed frame is a proper frame for the events, e_1 and e_2. In neither frame do the two events take place at the same location. In neither frame is the time interval between the events a proper time; the time dilation equation does not apply.

In the Rocket

We arrange for a light pulse to be emitted from the bullet at the point *a* in the rocket, at the time of event e_1, in such a way that it reflects off a mirror, *M*, and meets up with the bullet at point *b* at exactly the time of event e_2 (Fig 12.4). The mirror might have to be quite far from the path of the bullet, maybe even outside the rocket, as it is in Fig 12.4, since the speed of light most likely far exceeds the speed of the bullet. But this is no problem; it is only necessary that we can *imagine* such a mirror.

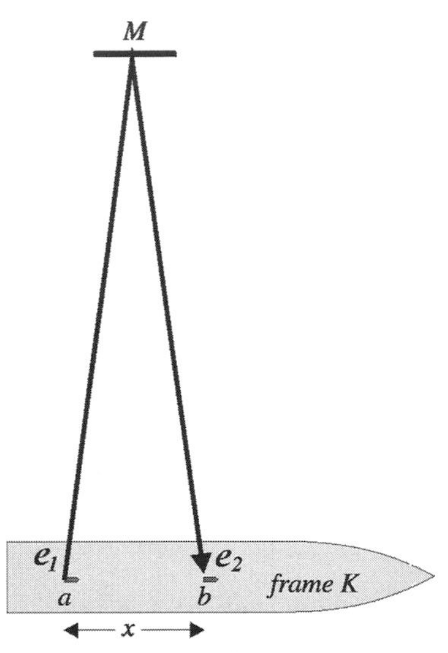

Fig 12.4 A light pulse is reflected off Mirror, *M*, and meets the bullet at *b*.

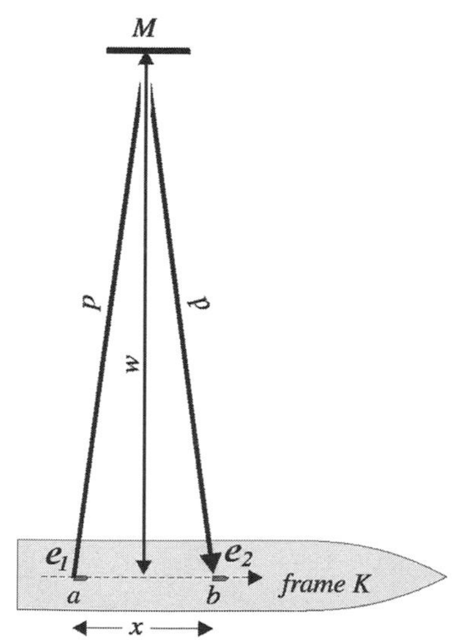

Fig 12.5 The light pulse takes time = $2p/c$ to reach point *b*.

The time, *t*, that it takes the bullet to go from *a* to *b*, and *x*, the distance *ab*, together will determine the perpendicular distance, *w*, which is how far from the bullet's path we have to place the mirror.

The total path length of the light pulse is $2p$. ("*p*" is a distance in the diagram, not "momentum.") The time of travel of the light pulse is, therefore, $2p/c$. The condition that the light pulse meets up with the bullet at *b* means that $2p/c = t$. This can be regarded as defining the distance, *w*.

p is the length of the hypotenuse of a right triangle with sides, w and ½x. By Pythagoras' theorem,

$$ct = \sqrt{4w^2 + x^2} \qquad [12.5a]$$

Outside the Rocket

To impose the condition that the speed of light in both frames of reference is *c*, we now perform a similar analysis of the very same light pulse, but as viewed from outside the rocket. Because the path is now longer, the time taken by the light pulse between the two events is longer, and t' will be longer than t. The analysis of the geometry of Fig 12.6 will give us an expression for t'.

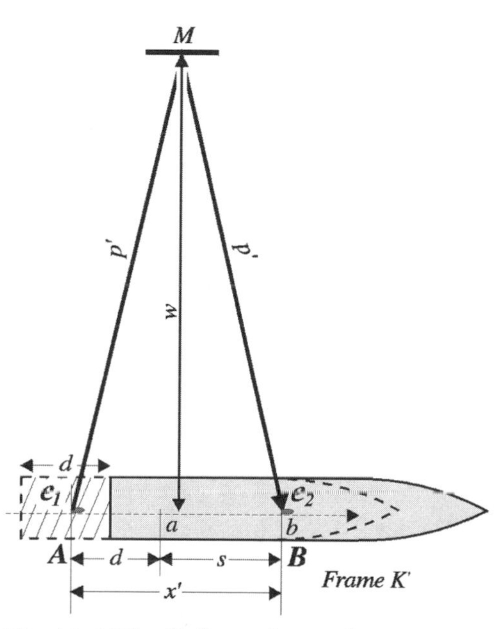

Fig 12.6 The light pulse path as seen in the fixed frame, *K'*.

The length of the path of the light pulse here is $2p'$, where p' is the hypotenuse of a triangle with sides, w and ½x', with w being the same as before, and x' to be determined. w is the same as it was in the rocket frame — it is, after all, the same light pulse and the same path. w is perpendicular to the direction of the rocket's motion, and so is not affected by the motion of the rocket frame.

The duration of the light pulse journey in the fixed frame, K', is $2p'/c$. This duration is equal to the time, t', between events e_1 and e_2 in the fixed frame. Pythagorean geometry produces an equation similar to [12.5a],

$$ct' = \sqrt{4w^2 + x'^2} \qquad [12.5b]$$

Squaring both sides of [12.5a], and doing the same to [12.5b] produces two simultaneous equations from which the variable, w, can be easily eliminated. The result is an equation with the two unknowns, t' and x'.

$$c^2 (t'^2 - t^2) = (x'^2 - x^2) \qquad [12.6]$$

Earlier in the chapter, it was found that

$$x' = ut' + x/\gamma$$
$$\text{where} \quad \gamma = 1/\sqrt{1 - u^2/c^2} \qquad [12.4]$$

The two equations, [12.6] and [12.4] form a pair of simultaneous equations. If t' and x' are unknown, these equations can be solved, giving these variables in terms of the given quantities, t, x, and u.

The solution is straightforward, if not entirely simple. Substitution of the expression in [12.4] for x' in [12.6] gives a single equation with just the one unknown, t'. This takes the quadratic form,

$$(c^2 - u^2)(t')^2 + (-2ux/\gamma)(t') + [-t^2c^2 + x^2(1 - 1/\gamma^2)] = 0 \qquad [12.7]$$

Using the expression for γ given in [12.4], simplification of the quadratic expansion is rapid, and gives the result,

$$t' = \gamma(t + ux/c^2) \qquad [12.8]$$

Substituting this result for t' back into [12.4] gives,

$$x' = \gamma(x + ut) \qquad [12.9]$$

The equations [12.8] and [12.9] are, for practical purposes, the equations of the Lorentz transformation. The standard form in which the Lorentz transformation is usually stated differs from these equations in notation and convention only.

The Standard Form of the Lorentz transformation

In the standard form of the Lorentz transformation, the velocity, u, is, by convention, the velocity of the prime (') frame with respect to the unprime frame. In the derivation of this chapter, u is the velocity of the unprime frame with respect to the prime (') frame, which was natural for the velocity addition problem. To bring equations [12.8] and [12.9] into the standard form, it is only necessary to reverse the sign of u, that is, replace u with $-u$, since the relative velocities are simply reciprocal.

In the standard form, also, the conventional symbol, β, is used to represent the ratio, u/c. Here, then, is the standard form:

The Lorentz Transformation

$$x' = \gamma(x - \beta ct) \qquad [12.10a]$$

$$ct' = \gamma(ct - \beta x) \qquad [12.10b]$$

and its inverse:

$$x = \gamma(x' + \beta ct') \qquad [12.10c]$$

$$ct = \gamma(ct' + \beta x') \qquad [12.10d]$$

u is the velocity of the prime (') frame with respect to the unprime frame
c is the velocity of light; $\beta = u/c$; and $\gamma = 1/\sqrt{1 - u^2/c^2}$

The astute reader will have observed that, in the derivation of the Lorentz transformation given above, it is assumed that there exists a proper time between events e_1 and e_2. This is implicit in using the light pulse clock; if, in the "rocket frame," the space separation between the events is great enough and the time small enough, there may be insufficient time for a light pulse to be at both events, and there is no proper time. The extreme example occurs when the events are simultaneous. This is not a limitation on the use of equations [12.8] and [12.9] in the next chapter, for in the velocity addition problem there is always a proper time.

Einstein's classic derivation of the Lorentz equations, avoids the present limitation, but is more abstract and difficult to follow. It appears in an appendix to his 1916 book, "Relativity", reprinted in translation by Crown Publishers.

Lorentz transformation Example 1: Time dilation

"Time dilation" is the name given to the transformation of a time interval between two events occurring at the same location in an "unprime" frame. In that frame, the time, Δt, between the events, is the shortest that it can be in any frame. This time is called the "proper time," and the frame is called the "proper frame."

The Lorentz transformation permits us to calculate, for the same two events, the time, $\Delta t'$, in a "prime" frame which is moving with respect to the proper frame at a speed v. Using equation [12.10b], and making intervals explicit with the symbol, Δ,

$$c\Delta t' = \gamma(c\Delta t - \beta\Delta x)$$

Because the two events are at the same location in the proper frame, $\Delta x = 0$. The time dilation equation follows immediately,

$$\Delta t' = \gamma\,\Delta t$$

The Phenomena of Relativity

13. The bullet in the rocket: Velocity Addition

We recognized earlier that the classical theorem of velocity addition (eq 3.2) is incorrect when the velocities are great.

That *classical* theorem of velocity addition says that if a bullet is fired with a velocity v toward the front of a rocket ship whose velocity with respect to a fixed frame of reference (the earth, or the solar system) is u, then the bullet's velocity, V, with respect to the fixed frame is $u + v$.

In the previous chapter we derived equations which would allow us to find the velocity addition theorem that is consistent with the postulates of relativity, rather than with the assumption of absolute time and fixed space.

We considered the problem (Fig 12.1) of the bullet fired with velocity, v, in a rocket whose velocity with respect to a fixed observer is u. The problem is to determine V, the velocity of the bullet with respect to the fixed observer, in terms of u and v.

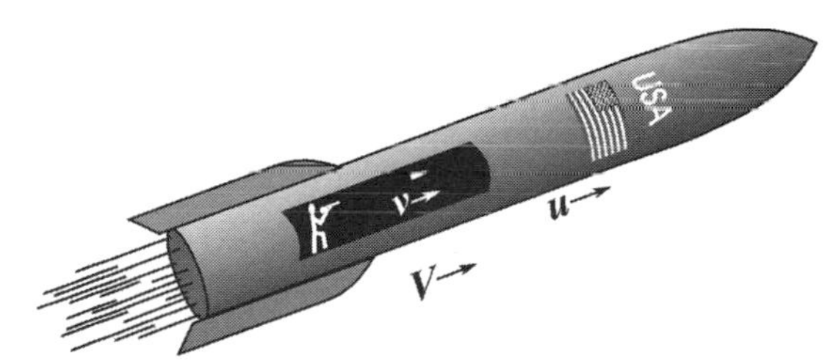

Fig 12.1 The rocket velocity is u. The bullet's velocity *in the rocket* is v. How do these velocities add?

We translated the problem into a form that led us to the Lorentz transformation. We supposed an interval of time, t, during which the bullet will have travelled a distance, x, with both x and t in the rocket frame, so that

$$x = vt \qquad [12.1]$$

Translated from velocities into times and displacements, the solution to the problem was the pair of equations,

$$t' = \gamma(t + ux/c^2) \qquad [12.8]$$

and

$$x' = \gamma(x + ut) \qquad [12.9]$$

where $\gamma = 1/\sqrt{1 - u^2/c^2}$

In the fixed frame, the "total" velocity of the bullet would be equal to the ratio, x'/t',

$$V = x'/t' \qquad [13.1]$$

This equation, with {12.8] and [12.9], gives,

$$V = \frac{x + ut}{t + ux/c^2} \qquad [13.2]$$

Substitution of the relation [12.1] into [13.2] gives the relativistic velocity addition equation,

$$V = \frac{u + v}{1 + vu/c^2} \qquad [13.3]$$

In the limit of small velocities, where $vu \ll c^2$, this equation reduces to the familiar Galilean rule, $V = u + v$.

If the "bullet" is a photon, whose velocity is the speed of light ($v = c$), [13.3] shows that no matter what the rocket speed, u, the speed V of the photon as observed by the fixed observer, is c.

If v is ½c, and u is also ½c, then the equation says that V is $0.8c$.

Example 1: Adding two velocities of 0.9c

v and u are each $0.9c$. What is V?

$$V = \frac{u + v}{1 + vu/c^2} \qquad [13.3]$$

$$V = \frac{0.9c + 0.9c}{1 + (0.9)^2} = 0.994c$$

Example 2:

You are walking at a speed of 3 m/sec down the aisle of a space ship whose speed relative to earth is $0.9c$. Show that an observer on earth would find that you are going only 0.57 m/s faster than the space ship. Would *you* feel that you were walking down the aisle at 3 m/s or 0.57 m/s? (Answer: 3 m/s)

The Phenomena of Relativity

14. Simultaneity

If two things happen at the same time, we refer to them as occurring "simultaneously."

Yet we know that the time that elapses between two events depends on who is keeping the time (in what frame of reference). Is it then possible that two events might be simultaneous in one frame and not in another? The very idea staggers the imagination, but the answer is, "Yes, that is possible, indeed!"

An often quoted example involves simultaneous lightning strikes at the front and rear ends of a moving railway car.

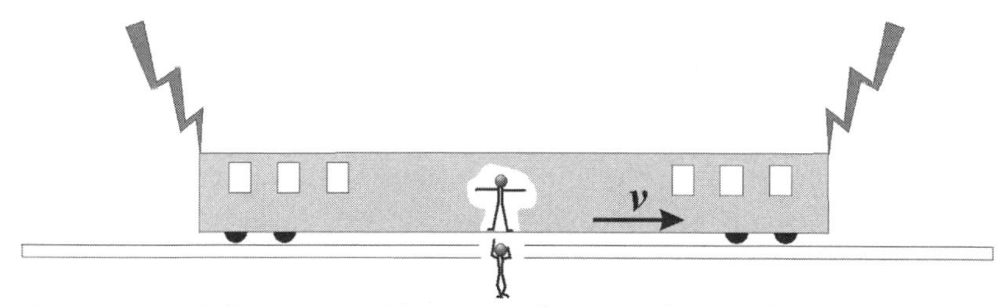

Fig 14.1 Lightning strikes simultaneously at the ends of the railway car. View is from the platform.

Let us assume that we are at rest on the platform, as the train passes by at great speed, v. Just as the mid-point of one of the railway cars is opposite where we are standing, two bolts of lightning strike the opposite ends of the car.

A moment later when the light from the two strikes reaches us, we reconstruct the events as follows:

The rest length, ℓ_o, of the railway car is contracted to ℓ_o/γ in our view. The strikes occurred half that distance to the left and to the right of us; we therefore see the strikes with a look-back time, $t_P = \frac{1}{2}\ell_o/(\gamma c)$, after they occurred ($t_P$ is "platform time;" neither the train nor the platform is a proper frame for the lightning strikes, which are space-separated in all frames). The fact that the flashes arrive at our location at the same time, having travelled equal distances, establishes that they occurred simultaneously (in our frame).

Fig 14.2 By the time the light beams from the two lightning strikes reach us, the car has moved a distance vt_P to the right.

We observe that when we see the flashes, the midpoint of the railway car is no longer directly in front of us, but has moved a distance $vt_P = \frac{1}{2}v\ell_o/(\gamma c)$ to the right. A passenger in the train standing in the aisle at the midpoint of the car will have already passed the light beam from the lightning strike at the front of the car, but the light beam from the strike at the rear will not yet have caught up with her. (The account of what is occurring inside the railway car is still being told in our frame, the platform.)

There is, then, an instant when, at least in our view, the passenger has already seen the flash from the front of the car, but could not yet have seen the flash from the rear.

The order in which two events occur at a single location may have physical consequences (as in the case of problem 5.1), and therefore can not depend on who observes them. It is therefore true also in the frame of reference of the moving train that there is an instant when light from the flash at the front has reached the passenger and light from the flash at the rear has not. The passenger, having never departed from the midpoint of the car, concludes that the flash at the front occurred first, since light from it has traversed the half-length of the car, $\frac{1}{2}\ell_o$, and arrived earlier than light from the flash that occurred an equal distance to the rear. The light from both flashes travels at the speed c in both the railway car frame and the platform frame.

The passenger, observing the strikes from her vantage point in the moving train, concludes that the strikes were not simultaneous.

In diagramming what happens inside the railway car, it would be a mistake to imply that the lightning flashes were simultaneous inside the car. The light from the flashes would travel, inside the car, exactly as it did outside the car. If one draws the flashes as simultaneous, what occurs inside the car must be portrayed from the point of view of the platform (Fig 14.3).

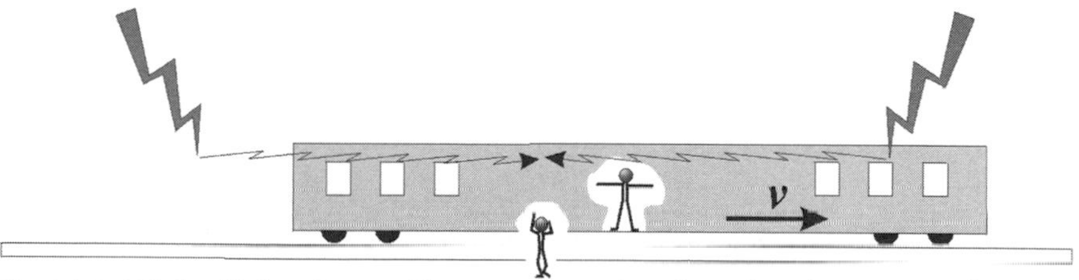

Fig 14.3 If the lightning strikes are portrayed as simultaneous, the view inside the railway car has to be in the platform (outside the car) frame.

> ***Note to reader:*** *The travel-time of the light flashes are important in this account, but their effects on the conclusion have been taken into account. The result of this thought experiment is therefore truly a relativistic effect.*

It is tempting, but not allowed, to ask such questions as, "In which frame did the lightning strike the front of the car first, or did it strike at the same time in both frames?" Clocks in ***two different frames*** can be synchronized at some instant at one location, but they will not remain synchronized. There is, in other words, no answer to the question, what time is it in your frame of reference when it is three o'clock in mine?

How Badly Out of Synch?

It is, however, legitimate to ask the following question: If two events are simultaneous in one frame of reference, how much time separates them in a different frame that moves at velocity v with respect to the first? In other words, how much later (in the frame of the railway car) does the light from the rear reach the passenger in the middle than the light from the front? Since the time required for light from the two flashes to reach the person in the middle is the same (in the frame of the railway car), this question is equivalent to asking, how much later (in the frame of the railway car) does lightning strike the front than the rear. This will obviously depend on the length of the car and the speed of the train.

Since neither frame is a proper frame for time between the two lightning strikes, the time dilation relation will not be helpful in translating time intervals. The Lorentz transformation would do this, but we have another tool that is equally useful: the invariant space-time interval (Chapter 9).

Since the two lightning strikes are simultaneous in the platform frame, the time separation between them in that frame, t_P, is zero. The flashes occur at the opposite ends of the railway car, which in the platform frame are separated by a distance $d_P = \ell_o/\gamma$.

The invariant space-time interval between the two events is therefore space-like, and is, according to [9.3b] (with $c^2 t_P^2 = 0$)

$$\sigma^2 \quad = \quad (\ell_o/\gamma)^2$$

In the train frame, the distance between the events is ℓ_o, and the time separation is the unknown t_T. The space-time interval in that frame is,

$$\sigma^2 = \ell_o^2 - c^2 t_T^2$$

The space-time interval is invariant with change of frame, so

$$(\ell_o/\gamma)^2 = \ell_o^2 - c^2 t_T^2$$

Since $\gamma = 1/\sqrt{1-v^2/c^2}$, with some algebra this becomes,

$$t_T = \ell_o v/c^2 \qquad [14.1]$$

This equation has the intuitively satisfying results at the extremes: $t_T = 0$ when $v = 0$; and $t_T = \ell_o/c$ when $v = c$.

The eleven-foot car in the ten-foot garage

The result in Eq 14.1 can help us understand the rather well-travelled puzzle of the "car in the garage." To be specific, let us choose a car that is 11 feet long (the indulgence of using feet as units rather than meters will not interfere), belonging to a person who has a garage that is only 10 feet long. It is obvious that in ordinary circumstances, the car will not fit in the garage (Fig 14.4).

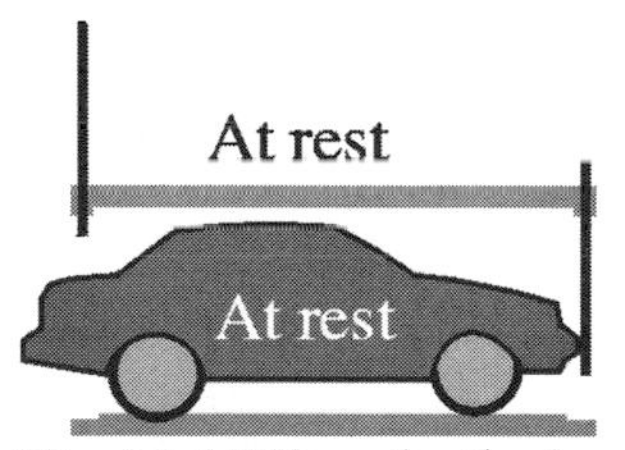

Fig 14.4 When both the garage and the car are at rest, the car does not fit in the garage.

If the car is moving rapidly, its length as observed from a position of rest, will be contracted, in accordance with the "length contraction equation," [8.3].

$$\ell = \ell_o/\gamma \qquad (\text{with } \gamma = 1/\sqrt{1-v^2/c^2}) \qquad [8.3]$$

Suppose that the car's speed is half the speed of light, $v = \frac{1}{2}c$, which makes $\gamma = 1.1547$. According to [8.3], then, the car's length is contracted to 9.526 feet, quite short enough to fit into the 10-foot garage. Of course, this shortening occurs only as long as the car is moving at a speed of $\frac{1}{2}c$, or 1.5×10^8 m/s.

Let us imagine that the garage has high-speed drop doors at both ends, as in Fig 14.5. If the car can travel at this enormous speed through the garage, there will be an instant during which it can be imagined that the doors are dropped simultaneously on both ends with the car inside, and there will be no collision of the car with either door as long as they are lifted back up quickly enough.

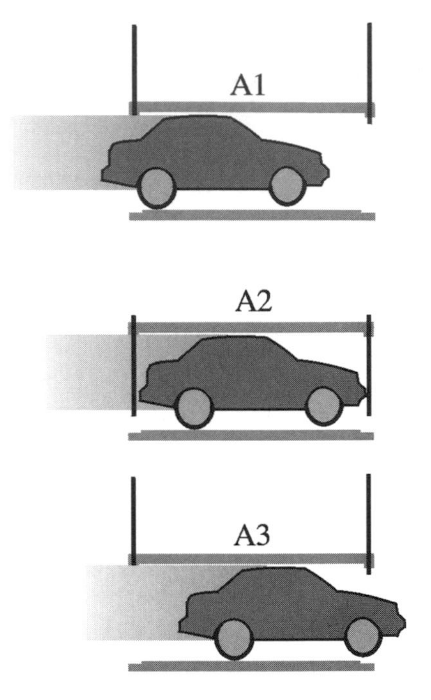

Fig 14.5 The shortened car can go through the garage and at one instant (A2) will be contained completely within it.

None of this is new to you. The examples in Chapter 8 have led you to expect this. What puzzles people initially is this question: How does this look from the point of view of the driver in this rapidly moving car. From the driver's point of view, the car is at rest, and the garage flies from right to left (in Fig 14.6) at the speed $v = \frac{1}{2}c$ with the consequence that the garage is shortened. In this frame of reference, the car, being at rest, of course, is not shortened. Since the garage is already shorter than the 11-foot car, its additional shortening by the factor γ gives the 10-foot garage a length, as viewed from the car, of only 8.660 feet. How will the driver be able to fit the 11-foot car into the 8.660-foot garage? After all, isn't there that instant when both doors can shut down on the car as in Fig 14.5? This can't *not happen* just because it is being observed by someone in a different frame of reference.

Of course, this is a simultaneity issue! Two events occur in that one instant: the front door drops down (event e_{front}) and the back door drops down (event e_{back}). These events are simultaneous, from the point of view of the stationary observer of Fig 14.5. But they need not be simultaneous in the frame of reference of the car.

Assume that, as shown in Fig 14.6, the front door drops (event e_{front}), and is quickly lifted again just before the front of the car would collide with it (B1).

As in the case of the railway car that is struck by lightning front and back simultaneously as seen from the platform, to the driver of the car the event e_{front} occurs by an amount of time, $t_T = \ell_o v/c^2 = 1.833 \times 10^{-8}$sec (Eq 14.1), earlier than the event e_{back}. Note that in Fig 14.6 the car is stationary, and the garage moves to the left, at the speed, $v = ½c$. In the time t_T the garage moves a distance $v t_T$, or 2.75 feet. Since the car (11 feet) is only 2.34 feet longer than the shortened garage (8.66 feet), the non-simultaneity is great enough to allow the garage to pass to the point where the rear door can be dropped behind the car (B3).

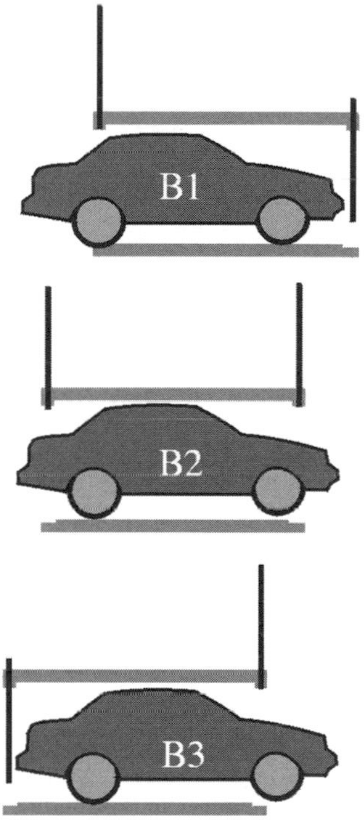

Fig 14.6 To the driver, the car is stationary, the garage whizzes by. The front door drops first, then the back door.

In each case, Fig 14.5 and Fig 14.6, the front door drops in front of the car, and the back door drops behind the car. But only to a stationary observer, as in Fig 14.5, do these two events occur simultaneously.

15. Transformations

Before we look more deeply at the Lorentz transformation, let us have a look at "transformations" in general. To transform means to change. As scientists observing the happenings of nature around us, we sometimes observe these happenings from different points of view. There may be various reasons for doing this. A new point of view may reveal something that is hidden in the old. Or, it may be that different observers are in fact viewing the same thing from different perspectives, not by choice, but by circumstance. In all of this, the one thing that does not change is nature itself, the "reality" out there being observed. (More on that later.) The *points of view* are usually referred to as *frames of reference*.

Surely the philosopher's tale of the two points of view regarding a glass half full of water ranks as the classic example of two ways of looking at the same fact: the pessimist says it is half empty, while the optimist says it is half full.

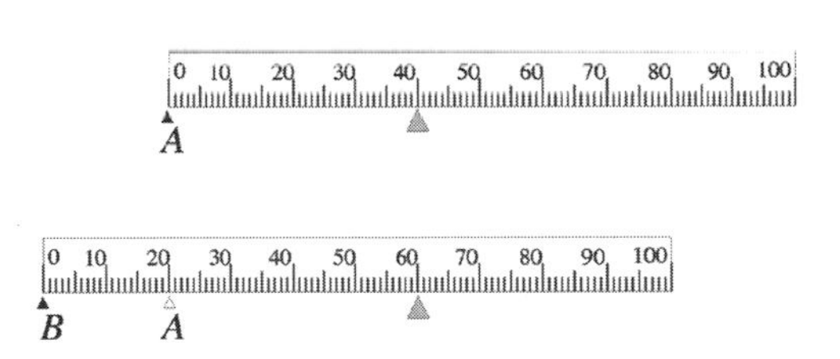

Fig 15.1 Transformation of variables: shift of the origin

A scientist places a meter stick with its zero next to a point A (Fig 15.1). An object is located next to the 40cm mark on the meter stick. With the origin of this one-dimensional frame of reference at A, the object is at x = 40cm. The meter stick is now moved so that its zero is next to a point B, 20cm to the left of A. The object is now next to the 60cm mark on the meter stick, at x' = 60cm. As a result of moving the origin 20cm to the left, the coordinates of all objects have undergone the transformation, $x' = x + 20\text{cm}$.

The location of the object has not changed. Only the value of its x-coordinate has changed as a result of moving the origin.

A historic transformation was proclaimed in 1543 when Nicholas Copernicus announced that, "finally we shall place the Sun himself at the center of the Universe." This changed the description of planetary orbits from the Ptolemaic system (with the earth as center), to the frame of reference with the sun at the center. Seen from earth, Mars has the characteristic retrograde (backing up) motion that objects moving in a circular path display when viewed from another object (earth) in a different circular orbit (Fig 15.2). With the path of Mars described in a frame of reference with the sun as center, its nearly circular orbit is revealed.

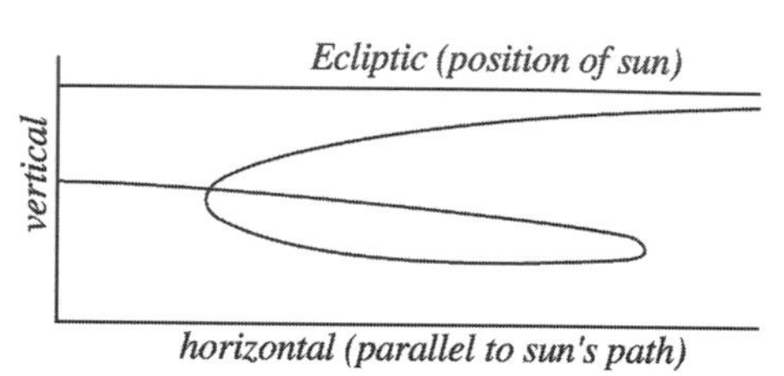

Fig 15.2 About 150 days in the path of Mars as seen from the earth.

This transformation is a mathematical operation, in itself neutral about the underlying physical events, which are represented equally *accurately* by both descriptions. The fact that one description is vastly *simpler* than the other suggests that it may be the one that portrays most nearly the way nature has designed it. From the Copernican, representation of planetary orbits, Newton was able to construct a law of gravitation that was a model of simplicity and that explained circular (and elliptical) orbits around the sun.

Changing the point of view

Transformations are changes in point of view. Observing physical phenomena from a variety of frames of reference, through a kind of tomography, taking "snap shots" from various angles, one gathers information about the phenomena themselves.

Frames of reference are commonly defined in terms of coordinate systems, involving measuring devices, such as meter sticks and clocks, to enable us to assign numbers to physical quantities such as location and time.

The physical quantities exist whether or not we have a way of assigning numbers to them. A point has location even in the absence of coordinates. Consider points, **P** and **Q**, in a plane (Fig 15.3). The points are *physical realities*. The picture shows where they are, but since there are no axes and no coordinates, we have *no way of describing* their location.

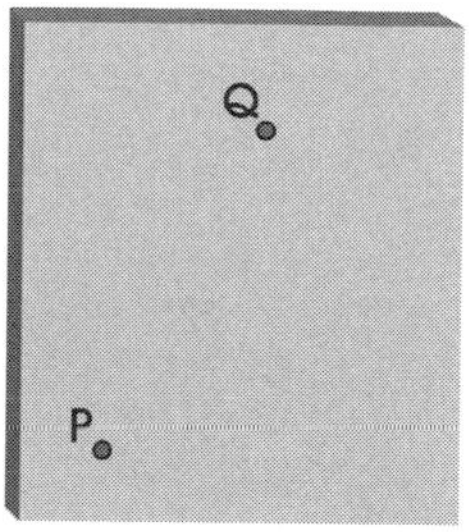

Fig 15.3 Points P and Q are physical realities.

To assign numbers that describe the locations of **P** and **Q**, we can imagine a glass plate with coordinate axes etched in it, slipped over the plane in which **P** and **Q** are located (Fig. 15.4). How we place the axes does not determine the location of the points; it determines the *numbers that describe their location*.

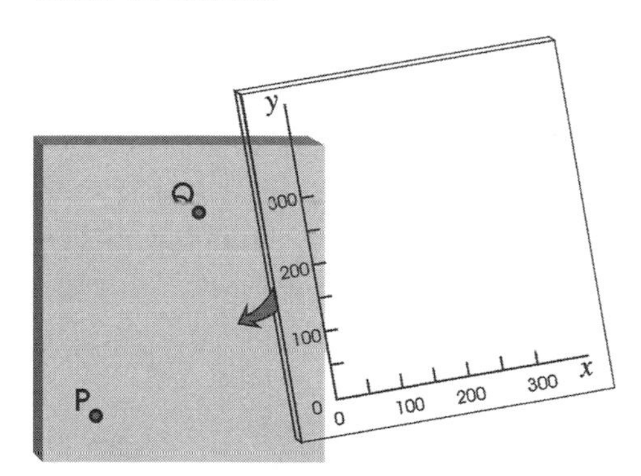

Fig 15.4 We can slip a glass plate with axes etched on it, over P and Q.

We can hold the glass plate any way we please. Suppose we choose to hold the glass plate with the x and y axes in the orientation shown in Fig 15.5, so that **P** is at the origin, ($x=0$; $y=0$). **Q** is then located at ($x=150$; $y=300$). The *interval*, **PQ**, between **P** and **Q**, is described as ($\Delta x=150$; $\Delta y=300$).

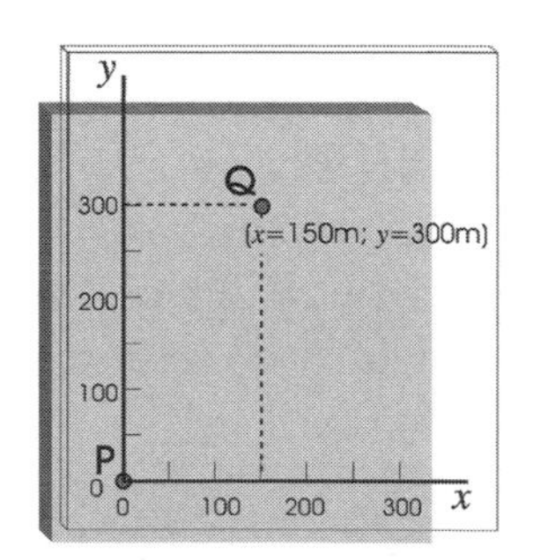

Fig 15.5 Axes have been placed over P and Q

In Fig 15.6, the glass plate has been rotated from its position in Fig 15.5, producing a new set of axes, (x',y'), that is every bit as good for describing the locations of the points as the set (x,y). To obtain the new set, the original axes are rotated around the origin until the y' axis passes through the point **Q**, whose coordinates are now $(x'=0;\ y'=335.4)$. The interval, **PQ**, is now $(\Delta x'=0;\ \Delta y'=335.4)$.

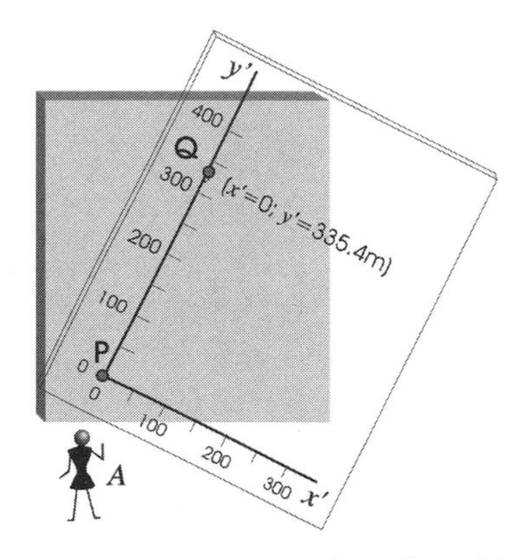

Fig 15.6 Axes can be placed in different ways over the same points, P and Q.

The change in the description of **PQ** from $(\Delta x=150;\ \Delta y=300)$ to $(\Delta x'=0;\ \Delta y'=335.4)$ is a *transformation from the x,y coordinate system to the x',y' coordinate system.* During the transformation, the *location* of **P** and **Q** remains unchanged; only the coordinate values change.

We have so far imagined the points **P** and **Q** seen by an observer *A* (oriented straight with the page of this book), who is standing still, rotating the glass plate, but *keeping her eyes and the page fixed.* Consequently in Figs 15.5 and 15.6 observer *A* sees **P** and **Q** fixed, while the axes are rotated.

Suppose that the points were viewed instead by an observer *B* whose orientation is fixed not to the page, but to the axes on the glass plate, as in Fig 15.6a. Believing, as we normally do, that "up" is defined as the "chin-to-nose" direction on our face, making "up" the direction of the y axis, observer *B* would conclude that the page has rotated, and that **Q** is now directly "above" **P**, as in Fig 15.7, which is simply Fig 15.6a rotated in its entirety to make "*B*" vertical.

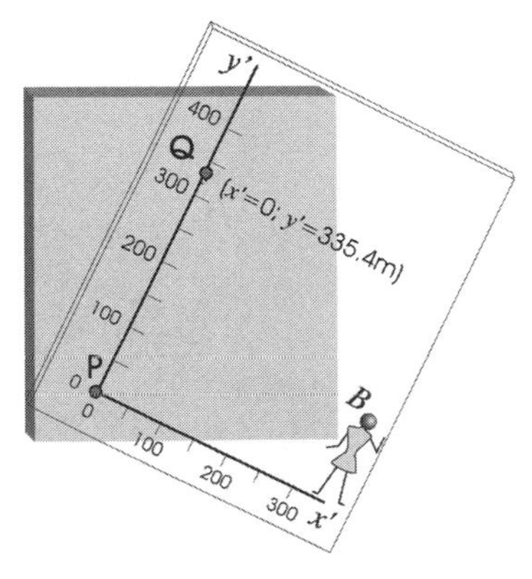

Fig 15.6a Observer *B* tilts at an angle, taking the glass plate with her.

To observer B the transformation of the frame of reference from (x,y) to (x',y'), appears as a rotation of the page and everything that is on it, from its original orientation in Fig 15.3 to that in Fig 15.7.

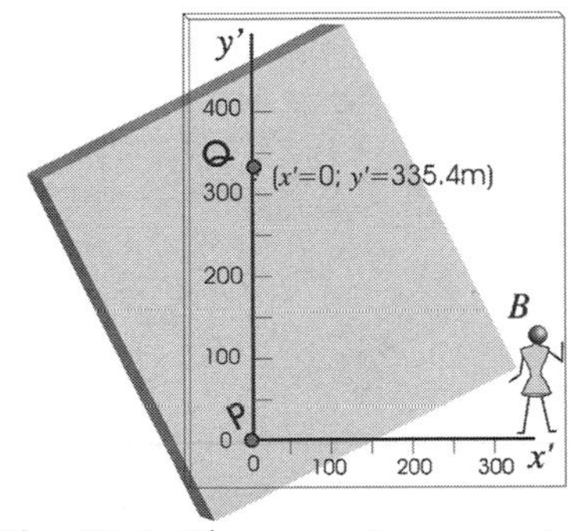

Fig 15.7 Observer B sees point Q directly "above" point P.

Were it not for external clues, such as a gravitational "vertical," or fixed objects in the surroundings, "B" might not be aware that her orientation has changed; her view of **P** and **Q** would be different from that of "A", even though **P** and **Q** have not changed their location or their orientation.

"A" believes that **PQ** is tilted about 26 degrees clockwise from *her* vertical (Figs 15.5 and 15.6). "B" believes that **PQ** is vertical, "chin-to-nose," (Fig 15.7). Who is right? What is the "reality" behind these two observations?

Observations and "physical reality"

If one rotates one's point of view from that of "A" (Fig 15.6) to that of "B (Fig 15.7), the orientation of the line **PQ** changes; whose description is better? Is there a "physical reality" beyond our various views of it? This is a truly cosmic question, which is not easily answered. The obvious fact is that the source of everything that we know about "reality" consists of our various views, or descriptions, of it.

David Mermin has dealt with this fascinating question as follows, "You and I may think that [science] works because, by a long and arduous process, scientists have become better and better at formulating questions that extract useful information from the natural world while avoiding questions that lead nowhere. This view is an expression of our naive realism, but it is important that we believe it. The conviction that we are trying to learn an objective truth is a powerful sustaining myth that drives us onward in our efforts..." [1]

Suppose that you walk around a telephone pole and photograph the pole from a dozen different angles. The observations (snap shots) from twelve different points of view represent twelve transformations of frame of reference. The twelve snap shots all look the same! "What a waste of effort!" you might say.

On the contrary, those twelve snapshots are evidence of a symmetry that characterizes the pole. You speculate that the pole has the shape of a cylinder. From your twelve observations, none of which alone allows this conclusion, you have drawn an inference about what you might describe as "the real nature" of the pole.

As an important complement to Mermin's observation, consider the following tip of the hat to the process of looking beyond the snap shots, from British astronomer Arthur Eddington, "If we are not content with the dull accumulation of experimental facts, if we make any deductions or generalizations, if we seek for any theory to guide us, some degree of speculation cannot be avoided."

What does *not* change as a result of a particular transformation can itself give important clues to the reality behind the observations. When Michelson and Morley found the speed of light to be the same in frames of reference of different velocity, they provided us with the equivalent of the twelve snap shots of the telephone pole. From their finding, with "some degree of speculation," the content of Einstein's special theory of relativity could be glimpsed.

More and more, physicists are finding that fundamental realities are defined by statements of invariance. These are called symmetries. The symmetry of the shape of the telephone pole is most elegantly described by the fact that its appearance is unchanged (is invariant) with rotation of the point of view.

---

[1] N. David Mermin, *Physics Today* **49 #3** March 1996 p11

Invariance in Euclidian Geometry and Relativity

We discovered in Chapter 9 that the space-time interval between two events remains unchanged when the events are observed in inertial frames of differing velocities. The square of the invariant space-time interval between two events, ($\Delta x^2 + \Delta y^2 + \Delta z^2 - c^2 \Delta t^2$), does not change when the point of view shifts.

This pseudo-pythagorean expression resembles the quantity that remains invariant as we rotate and shift the glass plate with the coordinate axes over the page containing the points **P** and **Q** in Figs 15.3–7. Even while changes occur in the coordinate values, x and y, and in the increments, Δx and Δy, the *magnitude of the space interval,* that is, the ***distance*** from **P** to **Q**, remains unchanged.

The distance, $|\mathbf{PQ}|$ is,

$$|\mathbf{PQ}| = \sqrt{(\Delta x)^2 + (\Delta y)^2} \qquad [15.1a]$$

The same calculation can be done in the (x',y') frame,

$$|\mathbf{PQ}|' = \sqrt{(\Delta x')^2 + (\Delta y')^2} \qquad [15.1b]$$

The result of these calculations is the same in both frames, $|\mathbf{PQ}|$ = 335.4m, and $|\mathbf{PQ}|'$ = 335.4m. It is easy to show that $|\mathbf{PQ}| = |\mathbf{PQ}|'$ with any other set of perpendicular axes, (x,y), in a plane surface in which the points **P** and **Q** are at rest.

In planar Euclidian geometry, the equations of transformation due to a rotation of perpendicular axes through the counter-clockwise angle θ are,

$$\begin{aligned} \Delta x' &= \Delta x \cos\theta + \Delta y \sin\theta \\ \Delta y' &= -\Delta x \sin\theta + \Delta y \cos\theta \end{aligned} \qquad [15.2]$$

From the transformation [15.2], one can derive the fact that the right hand sides of Eqs 15.1a and 15.1b are equal, and that |**PQ**| is invariant with transformations of rotation.

The *invariance* of |**PQ**|, Eq 15.1, is verification that the points **P** and **Q** exist in a "world" based on Euclidian geometry. This invariance establishes the "reality" of the Euclidian world, at least to the extent that our measurements are consistent with Eq 15.1.

A transformation can perform the function of testing, or verifying, the physical properties thought to underlie it.

A square post can not be distinguished from a round one in one snapshot, but the transformation rules for a square post predict that a sequence of 18 snapshots of it in silhouette, from angles that are 20° apart, will get first narrower and then wider four times, in a smooth way. Although it is not possible to compare the snapshots with the "real thing," one obtains verification through the consistency of observations from different angles with the transformation rules. These, in turn, are consequences of the postulated shape of the post.

The transformation [15.2], is based on the postulate that the geometry of the real world is Euclidian. From the transformation one can predict an invariance of |**PQ**| of the form [15.1]. An experimental verification of this invariance would then constitute evidence that the real world is Euclidian. By this same kind of reasoning, relativistic transformations are used to verify the postulates that underlie relativity.

Mapping *events* in classical space and time

The *points*, **P** and **Q** are locations in a plane.

Events, on the other hand, occur at certain locations at particular instants of time. If the point, **P**, were in motion so that at time t_o it is at (x_o,y_o), and at t_1 it is at (x_1,y_1), and at t_2 it is at (x_2,y_2), etc., then each of these events would be described by their "where" and their "when", by two space coordinates and one time coordinate: (x_o,y_o,t_o); (x_1,y_1,t_1); (x_2,y_2,t_2); etc.

The location of Mars in its orbit is a continuous sequence of events. Suppose we graph the *location* of Mars in a two dimensional space coordinate system in which the sun is at the origin. The orbit radius, R, is about 3,370 km; the planet makes one orbit in a period, T, of about 687 earth days. Two space coordinates, x and y lying in the plane of the orbit, and a time axis perpendicular to the x,y plane can be used to graph the sequence of *events* involved in Mars' progress through space as time elapses.

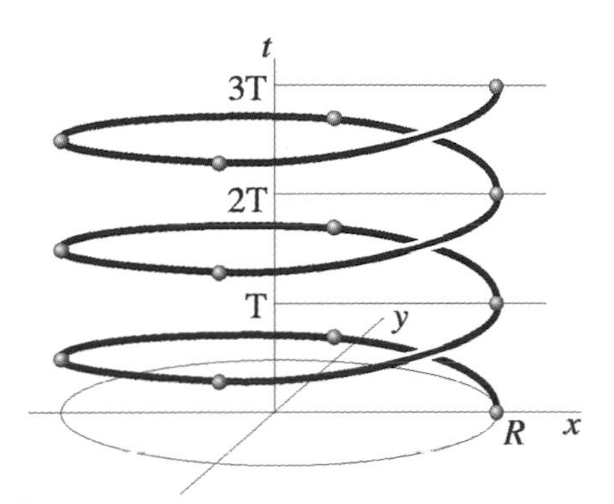

Fig 15.8 Mars' motion in orbit, graphed as a sequence of events in (x,y,t) coordinates.

If the zero of time is chosen when Mars' location is on the positive x-axis, ($y = 0$, $x = R$), then Mars' space coordinates will trace out a circle around the origin. Mars will return to the same point in the x-y plane, ($y = 0$, $x = R$), when $t = \mathrm{T}$ ($t = 687$ days), and again when $t = 2\mathrm{T}$, etc. In other words the events making up the history of Mars' location at various times form a spiral along the time axis.

It is clear from the discussion of the graphing of points **P** and **Q** that the space coordinates can be shifted and rotated. Moreover, with more drastic manipulation of the space coordinates, shifting the origin to remain fixed in the earth, the mapping of the path of Mars could be transformed into the Ptolemaic frame. Not a task anyone would want to do, but not impossible.

But what about the time axis? In the transformations described in the previous paragraph, the time axis has always remained intact, and vertical. Of course it has! These transformations have been made in the context of classical mechanics, where time elapses independently of how motion is represented in space. The first return of Mars to its original location (with respect to the sun) *has* to be at a time T after the beginning of the spiral. In classical space and time, a particular moment in time occurs at some absolute and universal value of the time variable, t, and at that moment of time the location of Mars is graphed at a particular point on a horizontal plane – a slice of space – that cuts the time axis at the height given by t.

Can relativistic space and time be mapped together?

Once we leave classical mechanics, and look for a geometry that can represent relativistic time and space together, what are the rules? Time and space do not proceed independently of each other. No longer are there neatly stacked, parallel horizontal slices, each a picture of where things are at some absolutely unique instant. The unfolding of the history of a sequence of events now depends on the velocity of the observer. The elapse of time on the observer's clock may not be synchronous with the elapse of time for a traveler on the moving object, any more than the earth observer's time is synchronous with that of the muon of chapter 7.

We know what the rules are. They have been developed in the previous chapters. Implementing them in a graphical description of a sequence of events is not as simple as rotating axes. Space and time are no longer absolute and distinct from each other. But space and time not *alike* in the way that North–South is like East–West, two directions that have different names but are measured in the same way.

Having looked at transformations of coordinates in x-y maps, we can accept that different numbers can describe the same locations. We are ready to consider the possibility that *time* can also be included in such maps, and that we can map the transformation of intervals between events, just as in classical space we mapped the transformation of intervals between locations. The transformation that we are looking to map is the Lorentz transformation.

16. Graphing the Lorentz transformation: Minkowski Diagrams

It is time to revisit the Lorentz transformation that was derived in Chapter 12, now that you have a good intuitive grasp of the mapping of simple transformations. The Lorentz transformation is a set of transformation equations like those of Chapter 15, except that, rather than dealing with rotation in ordinary (Euclidian) space, this transformation (Chapter 12) translates coordinate values in four dimensional (x,y,z,t) frames of reference.

A point in an (x,y,z,t) frame of reference is an event. The Lorentz transformation deals with the interval between two events $(\Delta x, \Delta y, \Delta z, \Delta t)$. Because it is understood that the values represent intervals, the symbol, "Δ" is usually omitted.

The Lorentz transformation is not difficult to use, and, as examples later in this chapter will show, can, in many cases, quickly and easily yield results obtainable only through lengthy analysis when done by the phenomenological methods of the earlier chapters.

Nevertheless, not all relativity problems fall into the category of those that the Lorentz transformation solves so readily. For example, length problems, and others that involve simultaneity in both frames, often do not yield readily to Lorentz solutions. The boat-and-bomb problem of Chapter 5 is likewise a problem not automatically solved by that transformation. As with many mathematical techniques, knowing when and how it applies can require greater insight than the application itself.

Prelude: the Minkowski diagram

Insight into an application is often clearest in graphical solutions; the Lorentz transformation can be performed graphically by the use of a "Minkowski diagram."

Representing four dimensions in a visual graph is virtually impossible; even three is difficult on two-dimensional paper. Fortunately, one loses no generality by laying out one of the axes - call it the x axis - along the direction of the relative motion of the two frames, so that motion in the y and z axes is unaffected, and can be disposed of by the identity transformations that require no graphing, $\Delta y' = \Delta y$ and $\Delta z' = \Delta z$. This conveniently leaves only the Δx component of the interval to be graphed against the time component, Δt .

Let us consider two events, e_1 and e_2 in a fixed frame of reference, (t,x), in which the observer is at rest. We can choose axes so that e_1 occurs at the origin — at $(t=0,\ x=0)$. Suppose that e_1 and e_2 are events in the life of an object moving in the $+x$ direction at a velocity ½c, half the speed of light. Let us say e_2 occurs one microsecond after e_1. The object will have travelled 150 meters. The coordinates of e_2 are:

$t = 1 \times 10^{-6}$ sec $\qquad x = 150$ m.

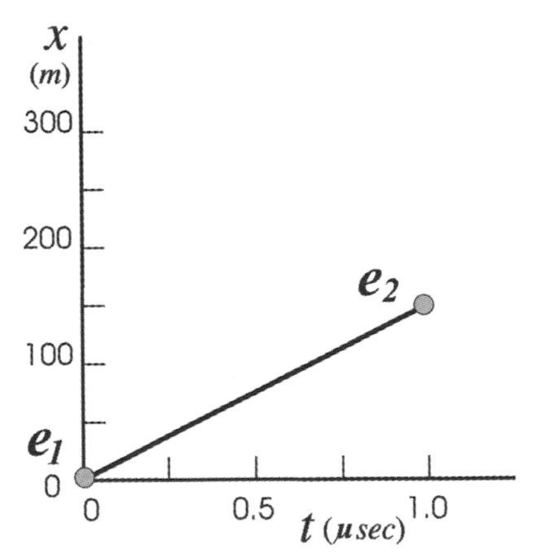

Fig 16.1 Two events in the life of an object moving at ½c.

The line between e_1 and e_2 (Fig 16.1) is called the *world line* of the interval e_1e_2. This phrase is used to avoid the misunderstanding that this line is a path in ordinary space; the world line always contains one component in the time dimension. The spiral of Fig 15.8 is the *world line* of the orbit of Mars.

The problem

We would like to know what the interval e_1e_2 looks like to an observer in a rocket moving, let us say, at a speed, $v_{\text{ROCKET}} = \frac{1}{4}c$ in the $+x$ direction, relative to the original frame of reference.

The premise is that the events are "physical realities," quite apart from their coordinates, in the same way that the two points **P** and **Q** were in Chapter 15. We are asking: What would the (t', x') axes look like, that would preserve the points that represent the events in their present place on the graph paper, but would give us their coordinates in the rocket frame?

The first difficulty is that, unlike the x, y plane, the t, x plane is not dimensionally uniform; t is measured in seconds, x in meters. What units would apply to an interval on a diagonal axis? Even relativity finds this an obstacle!

Recall the equations in chapter 9 for the space-time interval, σ^2,

$$\sigma^2 = \Delta x^2 + \Delta y^2 + \Delta z^2 - c^2 \Delta t^2 \quad [9.3b]$$

The component that is dimensionally compatible with Δx, Δy, and Δz, is evidently *not* Δt, but $c\Delta t$.[1] Using ct as the time variable gives us a measure of time that is *proportional* to *time* but has the units of distance. The quantity, ct, is measured in *meters*.

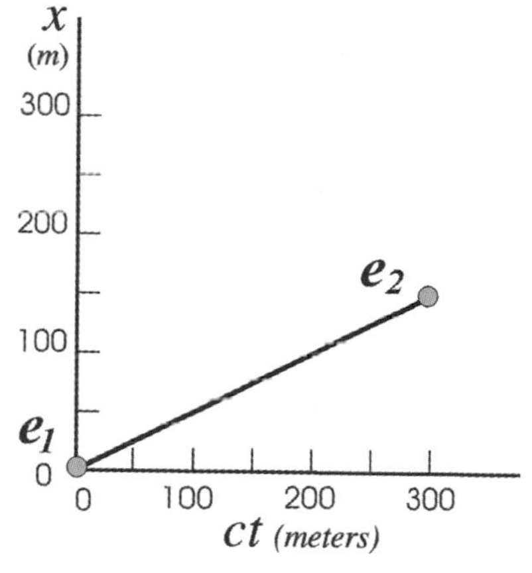

Fig 16.2 The events of Fig 16.1 graphed on a *ct* scale.

In Fig 16.2 the time axis in the graph of Fig 16.1 has been recalibrated, replacing, time units (seconds) with ct units (meters). When $t = 1\,\mu\text{sec}$, $ct = 300$ meters.[2]

---

[1] There is a minus sign, also, associated with the $c^2 t^2$ term (relative to the x^2... terms) that can not be totally ignored. It makes the ct variable imaginary (in the mathematical sense). The four variables of space-time are (x, y, z, ict) in four-dimensional space-time. But this need not concern us here.

[2] When $t = 1\mu\text{sec}$, $ct = (3 \times 10^8 \text{ meters/sec}) \times (1 \times 10^{-6} \text{ sec}) = 300$ meters

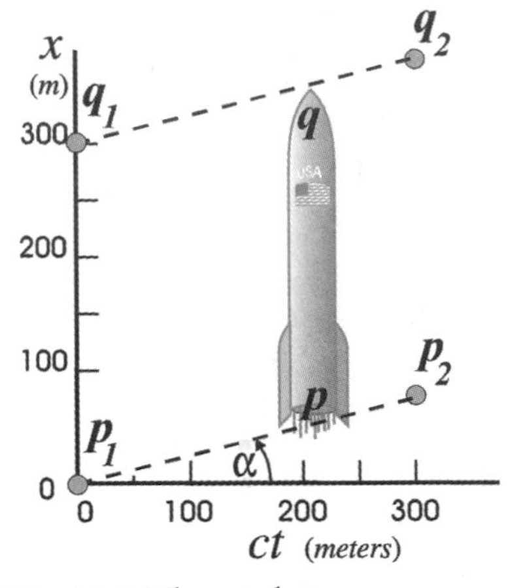

Fig 16.3 The rocket, *pq*, moves 75m along the *x* direction in 1μsec (*ct*=300m).

Let us temporarily remove from this graph the events e_1 and e_2, and instead display the motion of the rocket (Fig. 16.3). Label a point at the tail of the rocket *p*, and a point at the tip *q*. The rocket has a tail-to-tip rest length (as it would be measured by a passenger on board) of 309.8 m. The rocket's rest frame moves at a speed of ¼ *c* with respect to the fixed frame. At $v_{\text{ROCKET}} = \frac{1}{4}c$, γ is 1.033, making its contracted length, *pq*, in the fixed-frame *x* dimension equal to 300 m (see Eq 7.8b).

Fig 16.3 is a graph of the motion of the rocket in the fixed frame (ct, x), during an interval of 1 μsec ($c\Delta t = 300$m). The rocket, of course, does not move sideways; the horizontal axis is a time axis. The rocket moves only in the *x* direction, at constant speed, as the rocket frame must be an inertial frame. (This motion is not a launch, but perhaps a deep space voyage.) The dotted lines are world lines for the tail and tip of the rocket in the fixed frame.

At a speed of ¼*c*, the point *p* moves 75 meters in 1 μsec. p_1p_2 is the world line of the point *p* as it moves from p_1 ($x = 0, ct = 0$) to p_2 ($x = 75\text{m}, ct = 300\text{m}$). *q* moves at the same speed from q_1 to q_2.

The (contracted) rocket length *pq* remains the same (in the fixed frame) as the rocket moves. The world lines p_1p_2 and q_1q_2 are therefore parallel. The angle, α, that these world lines make with the *ct* axis of the fixed frame can be calculated from Fig 16.3.

The tangent of α is the slope of the world line, and is given by the ratio $\Delta x/c\Delta t$. The ratio, $\Delta x/\Delta t$ is the velocity, v, of the rocket; this makes $\tan\alpha = v/c$, and,

$$\alpha = \tan^{-1}(v/c) \qquad [16.1]$$

In the present example, where $v = \frac{1}{4}c$, this equation gives $\alpha = 14.04°$.

The time axis (ct') of the moving frame

The axes of Fig 16.3 are "fixed frame" axes, (ct, x), of an observer with respect to whom the rocket is travelling at $\frac{1}{4}c$. Now let us focus on the moving frame of reference, (ct', x'), in which the rocket is at rest. We are free to specify that the origin of the moving frame coincide with the origin of the fixed frame. Simply designate the time at which p is at p_1 to be the zero of t', as well as the zero of t; and designate the zero of the rocket axis, x', to be at the tail end of the rocket, at p_1, when $t' = 0$.

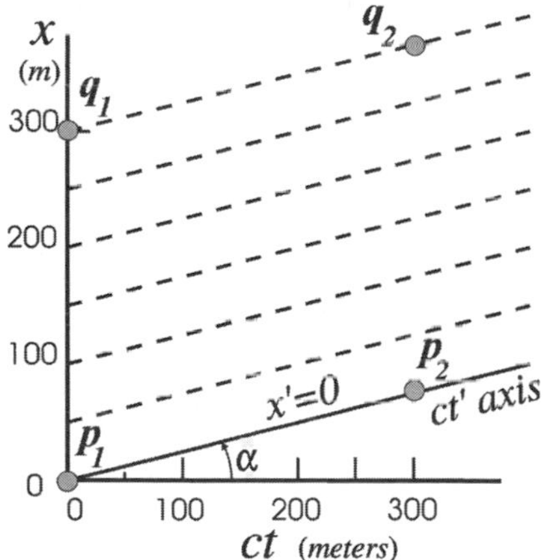

Fig 16.4 The time (ct') axis and other lines of constant x' in the moving frame of reference.

The clock in the moving frame, the "rocket clock," is started at the same time as the fixed clock, ($t' = 0$ and $t = 0$). At that instant the tail of the rocket ($x' = 0$), passes the origin of the fixed fame ($x = 0$).

In the frame of the rocket, the rocket does not move, so x' is zero along the world line p_1p_2 of the point p at the tail of the rocket. This makes a line through p_1 and p_2 the ct' axis, *the time axis of the moving frame*. The dotted lines drawn parallel to p_1p_2 are lines of constant x', or "ct' lines."

Having determined the orientation of the "ct' lines," which are lines of constant x', still leaves the orientation of the lines of constant ct', which are parallel to the x' axis to be resolved.

Search for the x' axis; lines of constant ct'

Looking at Fig 16.4, and recalling the rotated axes of chapter 15 (Fig 15.6), it is tempting to jump to the conclusion that the x' axis will be found perpendicular to the ct' axis, and that we can return to familiar ground: a pair of perpendicular axes rotated through a common angle. But this is not so. It is necessary to start afresh to determine what the x' axis looks like, by applying relativistic rules of physics, rather than precedents from classical physics.

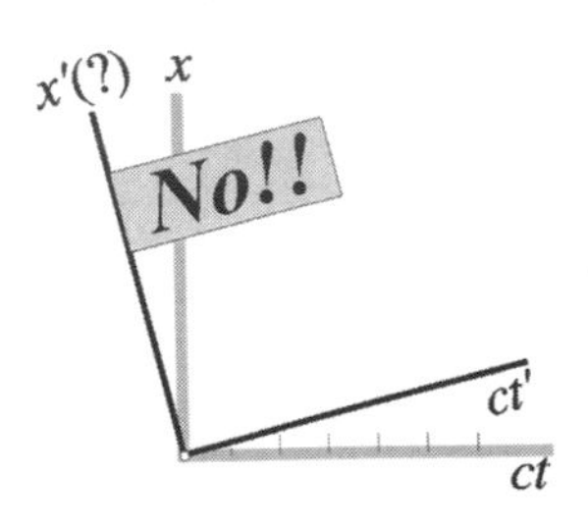

Are the axes rotated, as in Figs 15.6? NO! This diagram is wrong!

The x' axis is the line on which ct' is zero. Any two events that are simultaneous in the moving frame determine a line of constant ct'. Finding two such events and locating them on the graph of (Fig 16.4) will determine the direction of the x' axis. We know how to find two such points.

We know how to generate two simultaneous events in the rocket frame. A flash of light that occurs at the midpoint of the rocket, will send light pulses that will arrive at the tip and the tail of the rocket simultaneously (in the rocket frame), regardless of the motion of the rocket. Each of these pulses has to travel (in the rocket frame) half the (rest) length of the rocket. Travelling at the speed of light, c, the pulses will make the trip to the tip and to the tail in the same amount of time, arriving at p and at q at the same time in the rocket frame. Let us give the name p_3 to the event consisting of the arrival of the pulse at p, and the name q_3 to the arrival of the pulse at q.

To determine where the events p_3 and q_3 occur, we graph the world lines of the two light pulses in the fixed frame, superimposed on the world lines of p and q in the graph of Fig 16.4. Suppose the pulses originate at time $t = 0$ in the fixed frame, at a point, m, on the x axis of the diagram, half way between p_1 and q_1.

The pulses travel at the speed of light (in both directions) in the fixed frame as well as in the rocket — this is a principle of relativity! The velocities of the light pulse, $\Delta x/\Delta t$, are $+c$ and $-c$, making $\Delta x/c\Delta t = +1$ and -1. The *world lines* of the light pulses begin, therefore, at m and proceed with slopes of $+1$ and -1, as in Fig 16.5.

Fig 16.5 Lines from m are light pulses; dotted line p_3 to q_3 is a line of constant t'.

The events p_3, the arrival of one light pulse at the tail end of the rocket, and q_3, the arrival of the other light pulse at the tip of the rocket, occur where the world lines of the pulses cross the world lines of p and q. *In the fixed frame*, q_3 occurs later than p_3, which is not unexpected, because, in the fixed frame, the tail comes up to meet the pulse, while the tip runs away from it; *in the rocket frame* these events are simultaneous .

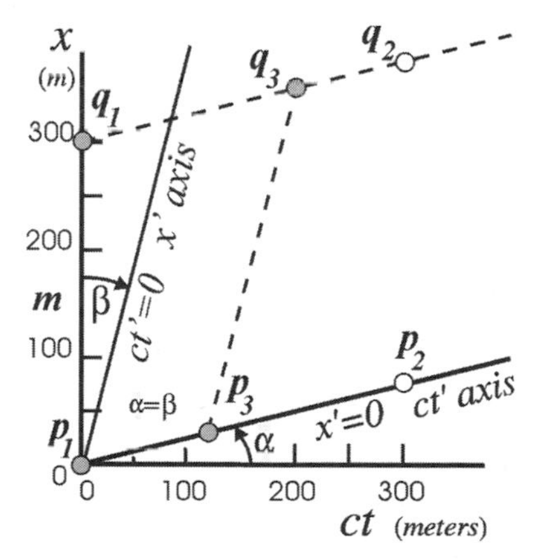

Fig 16.6 The x' axis is parallel to p_3q_3 and passes through the fixed frame's origin.

The dotted line, p_3q_3 is therefore a line of constant ct'. It was agreed earlier that the origin of the "rocket" frame is to pass through the origin of the fixed frame, so the line, p_3q_3 is not the x' axis; on this line ct' is constant, but not zero. The x' axis is therefore a line, parallel to the line p_3q_3, that passes through the common origin of both frames of reference (Fig 16.6). The x' axis makes an angle, temporarily called β, with the x axis. A geometric proof that the angles α and β are equal is given in Appendix II.

Coordinates in the rocket frame

Fig 16.7 is drawn from Fig 16.6, with the event points and constructions removed, to display the transformed axes (ct', x') of a frame of reference (the rocket frame) whose speed is ¼c with respect to the fixed frame.

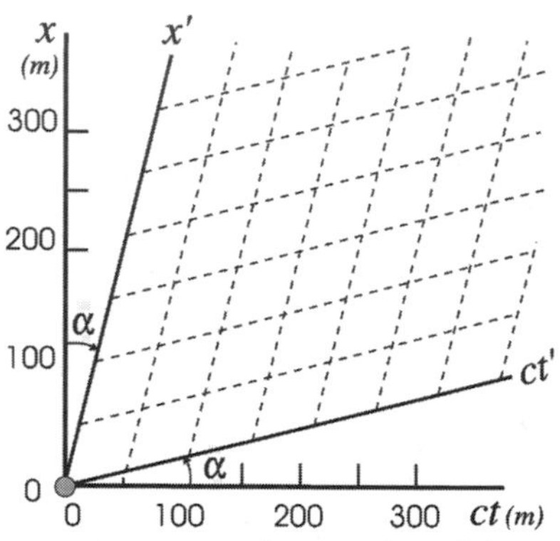

Fig 16.7 Coordinates in a frame moving at ¼c with respect to a fixed frame

The problem of scale remains. It can not be assumed that distances along the x and ct axes translate directly onto the x' and ct' axes.

We will define a scale factor, λ, as follows,

$$\lambda \quad = \quad \frac{\text{length of 1 unit in the moving (prime) frame}}{\text{length of 1 unit in the fixed (unprime) frame}} \qquad [16.2]$$

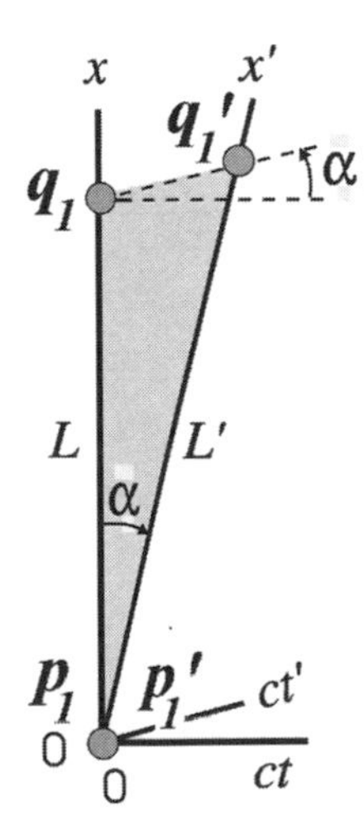

Fig 16.8 Part of Fig 16.6, to determine scale factor, λ.

The scale factor, λ can be found by comparing the lengths L and L' in Fig 16.8; this figure is a selected portion of Fig 16.6. The segment L represents the (contracted) length, $|p_1q_1|$, of the rocket in the fixed frame; L' represents the length, $|p_1'q_1'|$, of the rocket in its own frame, moving at ¼c with respect to the fixed frame.

Note that L is not equal to $|p_1q_1|$; nor is L' equal to $|p_1'q_1'|$.

$L/|p_1q_1|$ is the ratio of the length of the line in the diagram that represents $|p_1q_1|$ to the actual, real-life length of p_1q_1. It is the scale to which distances in the fixed frame are graphed.

$L'/|p_1'q_1'|$ is the ratio of the length of the line in the diagram that represents $|p_1'q_1'|$ to the actual, real-life length of $p_1'q_1'$. It is the scale to which distances in the rocket frame are graphed.

The ratio of these scales to each other is the ***scale factor***, λ.

The length contraction equation, [8.3] gives the ratio of the rocket lengths,

$$|p_1 q_1| = |p_1' q_1'| / \gamma \qquad [16.3]$$

(In the current example, $v = \frac{1}{4}c$, and $\gamma = 1/\sqrt{1 - v^2/c^2} = 1.033$.)

The ratio L'/L is obtained through a bit of manipulation from the geometry of the shaded triangle in Fig 16.8:

$$L'/L = \cos\alpha / \cos 2\alpha \qquad \text{from Fig 16.8}$$

$$|p_1' q_1'| / |p_1 q_1| = \gamma \qquad \text{from [16.3]}$$

Using the expression [16.1] for α, these relations give, for the scale factor,

$$\lambda = \frac{\sqrt{1 + v^2/c^2}}{\sqrt{1 - v^2/c^2}} \qquad [16.4]$$

The scale factor from a fixed frame to one whose relative speed is $\frac{1}{4}c$ is 1.0646. This means that along the x' axis of Fig 16.7, a 100 meter interval is 1.0646 times as long as a 100 meter interval along the x axis. This factor increases rapidly as the speed of the moving frame increases. When $v = \frac{1}{2}c$, the scale factor is 1.2910, and when $v = 0.9c$, it is 3.0865.

Fig 16.9 shows scaled axes, with bars to mark off 50 meter segments. The (x,ct) axes are for the fixed frame; the others are axes for frames whose velocities with respect to the fixed frame are as follows:

(x',ct') $v = \frac{1}{4}c$;

(x'',ct'') $v = \frac{1}{2}c$;

(x''',ct''') $v = 0.9c$;

(x'''',ct'''') $v = -\frac{1}{4}c$.

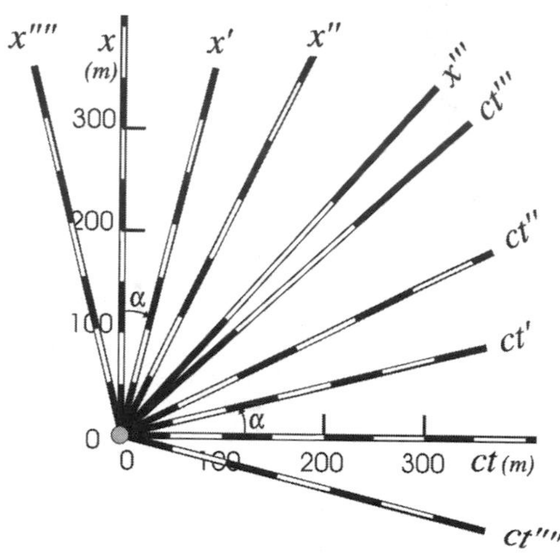

Fig 16.9 Axes for frames of various velocities with respect to a fixed frame, scaled in 50 meter segments.

Minkowski diagram Example 1: transforming an event interval

We now have the tools needed to transform the interval between the events e_1 and e_2 of Figs 16.1 and 16.2 from the fixed frame to a moving frame. A Minkowski diagram makes it possible to determine the distance and time intervals that an observer in the moving frame would find between the same events, e_1 and e_2.

In a "fixed" frame of reference, the events e_1 and e_2 occur 1 μsec apart in time. During this interval the object travels a distance of 150 meters, at a speed of $\frac{1}{2}c$ (see Fig 16.1). The time interval, 1×10^{-6}sec, has been translated into a ct interval of $(3 \times 10^{8}\text{m/s}) \times (1 \times 10^{-6}\text{sec})$, or 300 meters (see Fig 16.2).

To determine what the time and space intervals are for the same two events when they are observed from a frame of reference moving with a speed of $\frac{1}{4}c$ in the same direction as the object, we have placed the points representing e_1 and e_2 in a diagram

(Fig 16.10), that contains both the fixed frame axes and axes for a frame moving at $v = \frac{1}{4}c$, scaled by the factor λ.

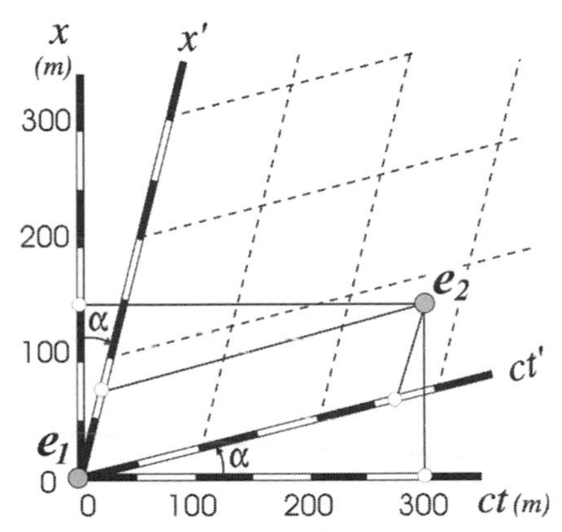

Fig 16.10 A Minkowski diagram showing components of e_1e_2 in two different frames of reference

Events e_1 and e_2 are from Fig 16.2. In the fixed frame, $\Delta x = 150\,\text{m}$, $\Delta ct = 300\text{m}$, and $\Delta t = 1.0\,\mu\text{sec}$. It is now possible to read from the diagram the coordinates one would observe in the moving frame. e_1 is at the origin of both frames; the coordinates of e_2 are the projections of e_2 onto the moving frame axes, ct' and x'. We find that $\Delta x' = 75\,\text{m}$, and that $\Delta ct' = 270\,\text{m}$. Dividing $\Delta ct'$ by c, we find that $\Delta t' = 0.90\,\mu\text{sec}$.

We can verify immediately that the space-time interval is the same in both frames. Because this space-time interval is time-like, τ^2 $(= c^2\Delta t^2 - \Delta x^2)$ is positive. A calculator finds this expression to be equal to $67{,}500\,\text{m}^2$ in the fixed frame. In the moving frame, the estimated readings from the Minkowski diagram, give the value, $67{,}275\,\text{m}^2$, for the expression $c^2(\Delta t')^2 - (\Delta x')^2$.

These transformations, as the example has shown, are ***not restricted*** to cases in which the reference frame is either a proper frame for time intervals, or a rest frame for space intervals. They can be used when the intervals between events have time and space components in ***both*** frames.

Minkowski diagram Example 2: A particle explodes

A particle at the origin of a fixed frame explodes at time $t = 0$ into two equal fragments which travel at speeds of $\frac{1}{2}c$ in opposite directions, $-x$ and $+x$. After $1\,\mu\text{sec}$ each fragment will have moved to a point 150 meters to either side of the origin. Refer to the *arrival of the fragments at these two locations* as events e_1 and e_2, respectively.

The two arrival events are simultaneous in the fixed frame, both occurring at $t = 1 \times 10^{-6}$ sec, that is, at $ct = 300$ m. The locations are $x_1 = -150$ m, $x_2 = +150$ m. In the fixed frame the fragments are 300 meters apart. How does this appear to a "traveler" on fragment 1?

In the fixed frame, the fragments separate a distance of 300m in 1 μsec. Does this mean that the speed of fragment 2 as seen by fragment 1 is equal to c? (The answer is "No.")

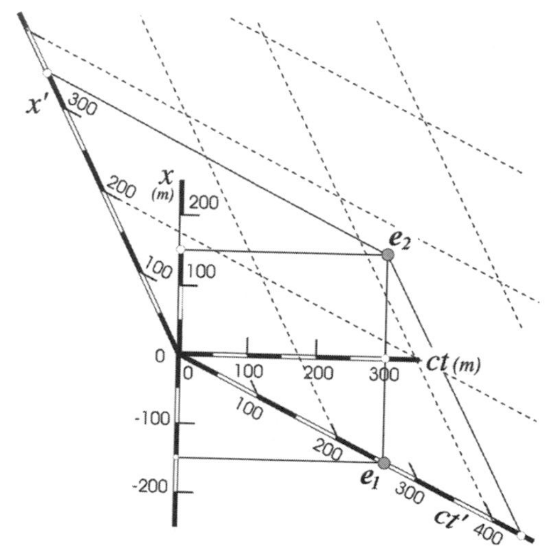

Fig 16.11 Minkowski diagram for the example of two particle fragments.

The problem can be solved using the Minkowski diagram of Fig 16.11. The (ct', x') frame is chosen to be the frame of an observer travelling with fragment 1. Its velocity with respect to the fixed frame is $-\frac{1}{2}c$. Because this velocity is negative, angle α is negative.

The event e_1 is situated on the moving frame's ct' axis, at:

$$x_1' = 0; \; ct_1' = 260\,\text{m}; \; t_1' = 260\,\text{m}/(3 \times 10^8\,\text{m/s}) = 0.867\,\mu\text{sec}$$

The projections of the event e_2 on the moving frame axes are:

$$x_2' = 345\,\text{m}; \; ct_2' = 430\,\text{m}; \; t_2' = 430\,\text{m}/(3 \times 10^8\,\text{m/s}) = 1.433\,\mu\text{sec}$$

What do these results tell us? Suppose that the problem arises as follows: there are two markers, 150 meters away, one in each direction from the point of the explosion.

An observer in the (fixed) frame, in which the source particle (the one that exploded) was at rest, would find that the fragments pass their respective markers, 300 meters apart, simultaneously.

Observers in the frame of reference of fragment 1 would clock fragment 1 passing its marker in 0.867 μsec, and fragment 2 passing its marker at $t_2' = 1.433\,\mu$sec (0.567 μsec after fragment 1 had passed its target). 345 meters is not the distance between the markers, but the distance from fragment 1, which has already passed its marker, to fragment 2 when fragment 2 passes its marker.

The question of interest here is about the ***velocity of fragment 2 as seen by fragment 1.*** This velocity is x_2'/t_2', or, $345\text{m}/1.433 \times 10^{-6}$ sec, which equals 2.408×10^8 m/sec, or $0.8025c$.

Minkowski diagram Example 3: Length contraction

A rod has length, ℓ_o, in a fixed frame in which it is at rest — its rest frame, the (ct,x) frame in the Minkowski diagram Fig 16.12. This length is measured by observing the ends of the rod simultaneously, let us say, at $ct = 0$. The events of this measurement are e_1 and e_2 in the diagram.

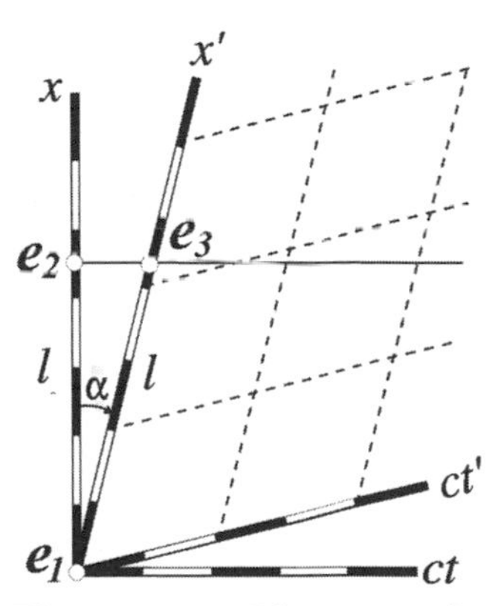

Fig 16.12 Diagram for transformation of lengths

It is desired to determine the length, ℓ, of this rod in a frame of reference that is moving at a speed, v, with respect to the rest frame. "Length," by definition, is the distance between two points observed simultaneously (Chap 8). The observations of the endpoints of the length, ℓ, have to be made simultaneously *in the moving frame*, meaning that these two observations have to be events on the x' axis, where $ct' = 0$.

In a conventional Lorentz transformation problem, the interval between two events is transformed from one frame to another. The events e_1 and e_2 of this problem are not simultaneous in the moving frame, (x',ct') – they are not both on the x' axis. The interval e_1e_2 is not a length measure in the moving frame.

The problem can be solved using the Minkowski diagram (or algebraically with the Lorentz transformation) only because the measurements of the ends, e_1 and e_2, of the rest length, ℓ_o, do not have to be made simultaneously. Because the rod is at rest, meter sticks can be laid next to it, end to end, at leisure, and read at any time. e_2 could therefore be at any point on a horizontal line $(x = \ell_o)$, including the point where this line crosses the x' axis. We label this point, e_3. The problem now is one of transforming the interval e_1e_3 from the fixed frame to the moving frame.

The length, ℓ, in the moving frame is therefore e_1e_3, measured along the x' axis. Using both the geometry of the triangle $e_1e_2e_3$, and the scale factor, λ, with Eqs 16.1 and 16.4, and the expression of Eq 7.8b for γ, gives the length contraction relation, Eq 8.3

$$\ell \;=\; (\ell_o / \lambda)(1/\cos\alpha) \;=\; \ell_o/\gamma$$

Summary of Minkowski diagram relations

Angles between fixed axes and moving frame axes

$$\alpha \;=\; \tan^{-1}(v/c) \qquad [16.1]$$

Scale factor

$$\lambda \;=\; \frac{\text{length of 1 unit in the moving frame}}{\text{length of 1 unit in the fixed frame}} \qquad [16.2]$$

$$\lambda \;=\; \frac{\sqrt{1 + v^2/c^2}}{\sqrt{1 - v^2/c^2}} \qquad [16.4]$$

where v is the velocity of the moving frame of reference
c is the speed of light

The Lorentz transformation

In the 1890's, Poincaré, Lorentz, Fitzgerald, and others, faced with the results of the Michelson-Morley experiment, described mathematically what they imagined as a shrinking of objects, resulting from rapid motion through the æther, causing an accompanying adjustment of time scales in the immediate surroundings. They found a mathematical solution that explained what they regarded as the artifact of the null result of Michelson and Morley's experiment.

The transformation they developed, still now called the Lorentz transformation, preceded Einstein's 1905 paper on Special Relativity. Lorentz acknowledged generously that it was Einstein who recognized that the mathematical result grows naturally out of the Maxwell equations when one discards the æther and abandons absolute time.

The Lorentz transformation is a more general relation than the equations for specialized phenomena such as time dilation and length contraction, and one must be cautious not to attempt to derive the general from the restricted.

The Minkowski diagram provides a graphical method of doing a Lorentz transformation between two frames of reference. It is quite as general as the Lorentz transformation. The only restriction is that both frames be inertial frames; this restriction applies equally to the graphical (Minkowski) and the algebraic (Lorentz) transformation. One can derive the Lorentz transformation from the Minkowski method, by expressing the geometry algebraically.

The classic derivation of the Lorentz equations, relying only on the postulates of relativity, is the one Albert Einstein called "Simple Derivation of the Lorentz Transformation," in Appendix I of his 1916 book, *Relativity* (available currently in reprint).[3] Although the derivation is rather abstract, and occupies six pages, it is understandable. The derivation presented in this book in

[3] A. Einstein, *Relativity*, 1916, 15th ed, translated by R.W.Lawson, Crown Publishers NY 1952, Appendix I pp 115–120

Chapter 12, using the device of the light pulse clock, is more grounded in tangibles. It makes the assumption that the events are time-like; another, similar, derivation would have to be constructed for the case of space-like events to make it completely general.

The algebraic Lorentz transformation makes the same assumption that was made in the Minkowski solution: that the four-dimensional problem can be reduced to two dimensions without loss of generality, by placing the x and x' directions along the direction of relative motion of the frames. The frames are inertial; the relative velocity of the frames is constant in both magnitude and direction. The Lorentz transformation is reconstituted to the four dimensional world by adding the two identity relations, $y' = y$; $z' = z$, meaning that the motion of the observers has no effect on displacements in the y and z directions.

The Lorentz transformation is a relation of time and distance *intervals* between the same pair of events, observed in two different frames of reference. Every t and every t' is really a time *interval*; similarly every x and x' is really an *interval* on the x or x' axis. The symbols "Δ" in Δt, $\Delta t'$, Δx, and $\Delta x'$ are typically omitted. The transformation is repeated here from Chapter 12.

The Lorentz Transformation

$$x' = \gamma(x - \beta ct) \qquad [12.10a]$$

$$ct' = \gamma(ct - \beta x) \qquad [12.10b]$$

and its inverse:

$$x = \gamma(x' + \beta ct') \qquad [12.10c]$$

$$ct = \gamma(ct' + \beta x') \qquad [12.10d]$$

u is the velocity of the prime (') frame with respect to the unprime frame

c is the velocity of light

$\beta = u/c$; and $\gamma = 1/\sqrt{1 - u^2/c^2}$

Lorentz transformation Example 1: Lightning strikes a railway car

In Chapter 14 we considered two lightning strikes that occur at opposite ends of a railway car. The strikes are simultaneous when observed in a "platform" frame. The railway car is in motion at a speed v with respect to the platform frame.

The question was, if two clocks at opposite ends of the railway car, previously synchronized in the moving train frame, are stopped as each is struck by the lightning, what difference in time would they show when later examined? In other words, how far out of simultaneity are the lightning strikes as seen on the train?

The events are the two lightning strikes. In the platform frame, designated as the "prime" frame, the strikes are simultaneous: the time between the events is zero,

$$\Delta t' = 0; \quad \text{and therefore} \quad c\Delta t' = 0 \qquad [16.6]$$

From the Lorentz transformation,

$$c\Delta t' = \gamma\,(c\Delta t - \beta\,\Delta x) \qquad [12.10b]$$

The left side of this equation is zero [16.6], therefore

$$(c\Delta t - \beta\,\Delta x) = 0 \quad or \quad c\Delta t = \beta\,\Delta x \qquad [16.7]$$

Δx is the distance between the lightning strikes in the train frame. The distance between the strikes, measured in the train frame, is the length of the railway car, notwithstanding the fact that the strikes are not simultaneous in the train. That distance is the rest length, ℓ_0, of the railway car. Substituting $\Delta x = \ell_0$ in [16.7], we get,

$$\Delta t = \beta\,\ell_0 / c$$

or, finally,

$$\Delta t = v\ell_0 / c^2 \qquad [16.8]$$

This is the answer to the question, and is the same as the result, Eq 14.1, in Chapter 14.

Example 2 Light and particles from a Supernova solved by graphical and analytic transformation

A supernova is an explosion of a large star in the last stage of its life cycle, producing a brilliant light in the sky that lasts a few days only. The supernova scatters element-rich material into the universe, supplying planets like ours with oxygen, iron, sulfur, copper, uranium, and all the other high atomic mass elements that stars like the sun lack.

Suppose that a supernova occurred in a star that was moving away from the earth at a relative speed of $0.2c$. The explosion sent light and subatomic particles hurtling toward earth.

In earth observer time (or better, solar system observer time, since earth has orbital motion), the light from this supernova arrived at earth 1000 years later (exactly 1000 years later, for the sake of argument). We would say that the star that exploded was 1000 light years from earth at the time it exploded.

The particles, travelling slower than the light, arrived later. The particles were emitted travelling at a speed of $0.9c$ *with respect to the supernova star and its remains.*

The information given allows us to describe the events of importance in "earth coordinates." The "earth" frame of reference contains an x axis that stretches from its zero at earth to the point where the supernova exploded 1000 LY (light years) away. The earth frame also contains a ct axis whose zero is the instant of the supernova explosion.

Event e_1, the emission of the light from the supernova, occurred at $ct = 0$, $x = 1000\text{LY}$.

Event e_2, the arrival of the light at earth, occurred 1000 years later, when t was 1000 years, ct was $(3 \times 10^8\,\text{m/s})(1000\,\text{years})$, the distance travelled by light in 1000 years. Event e_2 therefore occurred at $ct = 1000\,\text{LY}$, $x = 0$.

Event e_3, the emission of the particles from the supernova, occurred at the same time and location as e_1.

Event e_4, the arrival of the particles at earth, occurred at $x = 0$, at a time that can be calculated after first determining the velocity of the particles with respect to earth.

The Questions:

(A) find the velocity of the particles relative to earth (solar system), and from that result calculate the earth time coordinate, ct, of event e_4, the time of arrival of the particles on earth.
(B through E) Define a frame of reference, the "supernova frame," (ct', x'), moving at the speed of the supernova star (away from earth at a speed of $0.2c$). For convenience, let its origin (both time and space) coincide with the origin of the earth (or solar system) frame. In this frame find:

(B) the time of arrival of the light at earth ($\Delta t'$ of e_1e_2)

(C) distance traversed by the light to earth ($\Delta x'$ of e_1e_2)

(D) the time of arrival of the particles at earth ($\Delta t'$ of e_3e_4)

(E) distance traversed by the particles to earth ($\Delta x'$ of e_3e_4)

The Solutions:

(A) Use the velocity addition equation. u, the velocity of the moving frame is $+0.2c$. v, the velocity of the object relative to the moving frame is $-0.9c$. Eq 13.3 gives, $V = -0.854c$. The negative sign means the velocity is *toward* the earth; velocities away from the earth are in the positive x direction.

The time of arrival of the particles (in earth frame) will be their distance travelled (1000 LY) divided by their speed relative to earth: t = 1000LY/0.854c = 1171 years (the particles arrive 171 years after the light); and ct = 1000LY/0.854 = 1171 LY.

(B through E) can be solved graphically with a Minkowski diagram, or more exactly using the Lorentz transformation.

The Minkowski diagram (Fig 16.13) uses the earth frame as the "fixed" frame. The transformed axes represent the "supernova" frame, with a positive velocity (moving away from earth) of 0.2c. The angle between the axes, α, is 11.3° [Eq 16.1]; and the scale factor, λ, is 1.041 [Eq 16.4].

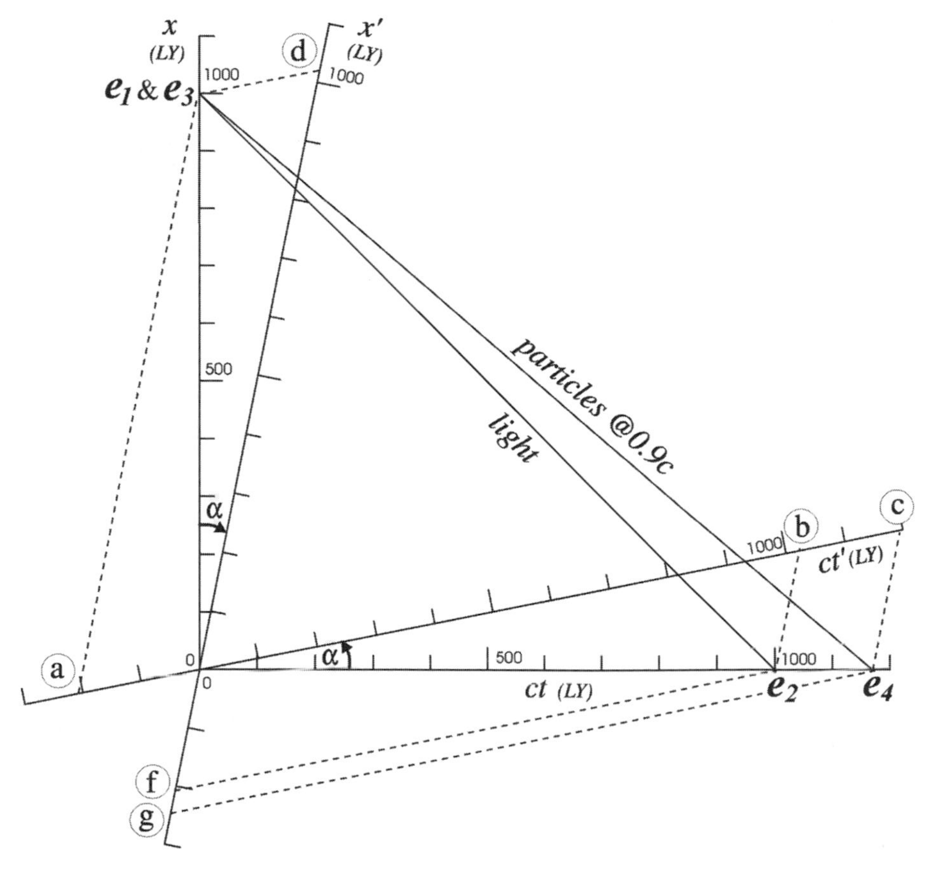

Fig 16.13 Minkowski diagram for light and particles emitted by a supernova 1000 light years away.

Using the Minkowski diagram to solve (B to E) graphically:

(B) $\Delta ct'$ of e_1e_2 is the interval between the projections of e_1 and e_2 on the ct' axis, between points labelled **a** and **b**: about 1240 LY. This makes $\Delta t'$ of e_1e_2 = 1240 years.

(C) $\Delta x'$ of e_1e_2 is the interval between the projections of e_1 and e_2 on the x' axis, between points labelled **d** and **f**: about 1240 LY. (Not surprising that light travels 1240 LY in 1240 years!)

(D) $\Delta ct'$ of e_3e_4 is the interval between the projections of e_3 and e_4 on the ct' axis, between points labelled **a** and **c**: about 1410 LY. This makes $\Delta t'$ of e_3e_4 = 1410 years.

(E) $\Delta x'$ of e_3e_4 is the interval between the projections of e_3 and e_4 on the x' axis, between points labelled **d** and **g**: about 1270 LY. (This gives a velocity in the supernova frame of 1270 LY/1410 Y or $0.901c$; this velocity was given as $0.9c$.)

Using the Lorentz Transformation to solve (B to E) algebraically:

This method involves solving the Lorentz transformation equations [12.10a] and [12.10b] for light and for particles. In all cases, $\beta = v/c = -0.2c/c$, or -0.2 (minus sign because v is directed opposite to the direction of motion of the objects - light and particles); and $\gamma = 1/\sqrt{1 - v^2/c^2} = 1.0206$.

(B) $c\Delta t' = \gamma(c\Delta t - \beta \Delta x)$ [12.10b]

Light from the supernova,

$c\Delta t$ = 1000LY, and Δx = 1000LY

$c\Delta t'$ = 1225 LY; $\Delta t'$ = 1225 years.

(C) $\Delta x' = \gamma(\Delta x - \beta c \Delta t)$ [12.10a]

$\Delta x' = 1225$ LY ; again, no surprise.

(D) $c\Delta t' = \gamma(c\Delta t - \beta \Delta x)$ [12.10b]

Particles from the supernova,

$c\Delta t = 1171\text{LY}, \quad \Delta x = 1000\text{LY}$

$c\Delta t' = 1399\,\text{LY}; \; \Delta t' = 1399$ years.

(E) $\Delta x' = \gamma(\Delta x - \beta c \Delta t)$ [12.10a]

$\Delta x' = 1260$ LY

17. Equations of Motion in Four Dimensional Spacetime

Before bringing this book to a close, we would like you to have a glimpse of what might be in store for readers who want to pursue this subject further: the use of relativistic equations of motion for solving particle interactions using momentum and energy concepts defined in the four dimensions of space-time.

In classical physics, Newton's Laws describe the interactions of objects through the concepts of mass, Force, and acceleration. From these laws one can derive the Work-Energy Theorem, and the Impulse-Momentum relation, and the classical conservation laws for energy and momentum.

Here we introduce equations that incorporate the time-space interrelation of special relativity in a system of conservation laws in the language of four-dimensional energy-momentum quantities. A few examples will show its power in solving problems of particle interactions, a major area of work in current physics.

In Chapters 10 and 11 we gave a rationale for definitions in relativistic mechanics for some key quantities: momentum, rest energy, kinetic energy, and total energy, in terms of rest mass, velocity, and the transformation variable γ.

Momentum: $\mathbf{p} = \gamma m_o \mathbf{v} = m'\mathbf{v}$ [11.1]

Rest energy: $E_o = m_o c^2$ [11.6]

Total energy: $E = \gamma m_o c^2 = m' c^2$ [11.7]

Kinetic energy: $K = E - E_o = (\gamma - 1) m_o c^2$ [11.8]

In the course of writing these definitions, it became convenient to think of the product, γm_o, as the "relativistic mass," m'.

Although the concept of "Force" is not as key a concept in relativistic mechanics as it is classical physics, Force has a definition which, in analogy to classical relations, defines it in terms of "that which causes momentum to change,"

Force: $$\mathbf{F} = \frac{d\mathbf{p}}{dt} \qquad [11.2a]$$

A system removed from all external influences obeys a conservation rule for momentum. These external influences are described more commonly, in modern physics, as "interactions," rather than forces.

"Work" is now defined in terms of (relativistic) momentum,

Work: $$dW = \mathbf{v} \cdot d\mathbf{p}$$

which leads to the familiar relation of the Work–Energy theorem, from which it can be concluded that a system upon which no net Work is done conserves total energy.

If the magnitude of **p** is designated by the scalar notation, p, then $p^2 = \mathbf{p} \cdot \mathbf{p}$, just as $v^2 = \mathbf{v} \cdot \mathbf{v}$. This means that, for a single particle or object, p^2 may be written:

$$p^2 = \gamma^2 m_o^2 v^2$$

Since $E_o^2 = m_o^2 c^4$ and $E^2 = \gamma^2 m_o^2 c^4$, it follows in a few lines of algebra that,

$$E^2 = E_o^2 + c^2 p^2$$

It becomes convenient for the purpose of notation, as well as for laying the groundwork for more advanced concepts, to define a new three dimensional vector quantity, also called "momentum," and differentiated from the earlier-defined momentum, **p**, by the use of the capital letter, **P**.

$$\mathbf{P} = c\mathbf{p} \qquad [17.1]$$

from which it follows that if $P^2 = \mathbf{P}\cdot\mathbf{P}$, then $P^2 = c^2 p^2$, and

$$E^2 = E_o^{\,2} + P^2 \qquad [17.2]$$

The factor, c, in Eq [17.1] makes the vector **P** dimensionally an energy quantity. **P** is related to **p** simply by the factor, c, and must therefore obey any conservation law that is obeyed by **p**.

If now P is expanded according to Pythagoras' rule, Eq [17.2] becomes,

$$E^2 = P_x^{\,2} + P_y^{\,2} + P_z^{\,2} + E_o^{\,2} \qquad [17.3]$$

where the energy of the rest mass, E_o, appears as a fourth Pythagorean component of a four-dimensional vector whose magnitude is E, the total energy. The total energy may be regarded as the four-dimensional vector,

$$\mathbf{E} = \{P_x, P_y, P_z, E_o\} \qquad [17.4]$$

Relativistic Laws of Motion

With the notation defined above, the laws of motion for an "isolated" system of particles or objects, undergoing interactions among themselves only, with no influences from outside, can be written,

$$[\Sigma \mathbf{P}]_{\text{BEFORE}} = [\Sigma \mathbf{P}]_{\text{AFTER}} \quad [17.5]$$

$$[\Sigma \text{E}]_{\text{BEFORE}} = [\Sigma \text{E}]_{\text{AFTER}} \quad [17.6]$$

where the summation, Σ, refers to summation over all particles or objects in the system. "Before" and "after" refer to a given interaction, or sequence of interactions within the system.

Total momentum is conserved. Total energy is conserved. Kinetic Energy is *not* conserved, nor is mass, in the usual sense of particle rest mass. Another way to write [17.6] is,

$$[\Sigma (\text{K} + \text{E}_o)]_{\text{BEFORE}} = [\Sigma (\text{K} + \text{E}_o)]_{\text{AFTER}} \quad [17.7]$$

E_o is the energy of the rest mass of the particle. E_o can be called the "rest energy." Eq [17.7] implies that in an interaction there can be an interchange between rest mass and Kinetic Energy.

Particle creation, the formation of particles out of Kinetic Energy, and particle annihilation with the release of Kinetic Energy, are thus permitted. A somewhat wider view of Kinetic Energy is required at this point, to include the energy of so-called "massless particles," those with zero rest mass. These are particles such as photons, which travel at the speed of light, *c*, have momentum as well as energy, but whose rest energy, E_o, is zero.

Also included are the energies of the particles that make up the energy of fields, such as the electromagnetic field. The energy density (energy per cubic meter of space) of the electric field, for example, has a quite definite value, and can be calculated, in Joules/m³, from the equation, $u_E = \frac{1}{2}\epsilon_0 E^2$, with E the magnitude of the electric field and ϵ_0 the dielectric constant of free space. Because of the energy in the field, a certain amount of mass density is associated with it. The mass of the field contributes inertia as well as gravitational weight. In the case of a pair of charges, one positive and one negative, as for example a Chloride ion and a Sodium ion, the mass of the Sodium Chloride molecule is diminished by the decrease of the field due the formation of the dipole; the total mass of the dipole is less than the sum of the masses of the individual ions. The decrease is associated with the loss of potential (binding) energy.

For massless particles,

$$E = P$$

Quantum theory tells us that the energy, E_λ, of a photon is equal to the product of a universal constant, "h", called *Planck's constant*, and the frequency, υ, of the electromagnetic waves in the photon,

$$E_\lambda = h\upsilon$$

Planck's constant has the value: $h = 6.626 \times 10^{-34}$J sec. The frequency is related to the wave length, λ, by the relation,

$$\upsilon = c/\lambda$$

and so,

$$E_\lambda = hc/\lambda$$

from which it follows that the "small letter" momentum, p, of a photon is

$$p = P/c = E/c = h/\lambda$$

In particular, collisions between particles and anti-particles (particles of anti-matter) can be made to occur for the purpose of turning all the rest mass of both (plus whatever kinetic energy the colliding particles may possess) into massless particles of enormous energy. The inverse process also occurs; in the very early universe, the first particles are thought to have formed spontaneously from the "freezing" of pure energy.

Example 1: Positron – electron annihilation

Suppose a positron (an anti-electron) and an electron, neither moving very rapidly, annihilate, decaying into two photons of equal energy moving in opposite directions (if a single photon were produced, it could not conserve momentum). Find the wave length of the photons.

$$[\Sigma E]_{BEFORE} = [\Sigma E]_{AFTER}$$

$$2m_o c^2 = 2(hc/\lambda)$$

$$\lambda = h/m_o c$$

$$\lambda = (6.626\times 10^{-34}\text{J sec})/(9.11\times 10^{-31}\text{kg})(3\times 10^{8}\text{m/s})$$

$$\lambda = 2.4\times 10^{-12}\,\text{m}$$

A photon of this wave length is an extremely high-energy photon, comparable with those that make up high energy cosmic rays.

Example 2: Proton – antiproton pair production[1]

It is desired to blast a stationary proton (p^+) with another proton whose kinetic energy is so great that out of the collision will come not only the two original protons, but in addition also a proton(p^+)–antiproton(p^-) pair.

This reaction,

$$p^+ + p^+ \rightarrow p^+ + p^+ + p^+ + p^-$$

is possible if sufficient energy is provided, because electric charge and baryon (proton/neutron) number are both conserved (the anti-proton baryon number is –1). The rest mass of the new particles, the (p^+p^-) pair, has to be provided by the kinetic energy of the accelerated proton.

The question is, how much energy must the accelerated proton have, and what is its speed at that energy?

It is not determined how the momentum is to be distributed among the four product particles. A reasonable assumption is that the particles move out of the collision more or less together; this assumption gives us the right to assign to each of the four product particles the same total energy and the same momentum.

Labelling the energy and momentum of the original two particles with the subscripts "1" for the incident, or high-energy, proton, and "2" for the at-rest, or target, proton ($P_2 = 0$), and designating with a prime (') all particles after the collision, the conservation conditions become,

Conservation of Energy:

$$E_1 \quad + \quad E_2 \quad = \quad 4E'$$

$$\sqrt{E_{o,1}^2 + P_1^2} \quad + \quad E_{o,2} \quad = \quad 4\sqrt{E_o'^2 + P'^2}$$

[1] This example is from notes by Professor Fred Handel of the University of Michigan.

The rest masses, $E_{o,1}$, $E_{o,2}$, and E_o', of all the protons (including the anti-proton) are the same, and can be given a common symbol, E_o. Squaring both sides gives,

$$E_o^2 + P_1^2 + E_o^2 + 2E_o\sqrt{E_o^2 + P_1^2} = 16\,(E_o^2 + P'^2)$$

Conservation of Momentum:

$$\mathbf{P}_1 + 0 = 4\mathbf{P}'$$

Consequently, $P_1 = 4P'$, and therefore $P_1^2 = 16P'^2$. Substituting this in the last of the energy equations, above, gives,

$$2E_o\sqrt{E_o^2 + P_1^2} = 14E_o^2$$

The total energy, E_1, of the incident proton is just $\sqrt{E_o^2 + P_1^2}$, so

$$E_1 = 7E_o$$

$$K = E_1 - E_o = 6E_o$$

The rest energy, E_o of a proton is,

$$E_o = m_pc^2 = (1.67\times10^{-27}\text{kg})(3\times10^8\text{m/s})^2$$

$$E_o = 1.503\times10^{-10}\text{Joules} = 9.39\times10^8 \text{ electron volts}$$

The Kinetic Energy of the accelerated protons required for this experiment is therefore, six times 9.39×10^8 electron volts, or

$$K = 5.63\times10^9 \text{ electron volts}$$

(An electron volt is a unit of energy: the energy of an electron or proton accelerated through a voltage difference of one volt.)

An accelerator called a "bevatron" which can produce protons with kinetic energy of up to 6×10^9 eV was in fact used to produce the first artificial proton-antiproton pairs. A Nobel Prize was awarded for this achievement.

Protons of the required energy have a γ of 7, and are thus travelling at speeds of $0.990c$.

More Problems

1. Time Dilation

A π^+ meson has a half-life of 18 nanoseconds ($1\,\text{nsec} = 1 \times 10^{-9}\text{sec}$). This half-life is measured, as always, in the π^+ meson's own frame of reference. In the Penn-Princeton proton synchrotron, π^+ mesons are produced which travel at a speed of $0.9975c$.

(A) How far would a beam of these π^+ mesons travel before half of them dissociated, according to classical, Newtonian mechanics?

(B) What fraction of the original π^+ mesons would remain in the beam a distance of 160 meters from the point of origin, according to classical Newtonian mechanics?

(C) How far would a beam of these π^+ mesons *actually* travel before half of them dissociated?

(D) What fraction of the original π^+ mesons would *actually* remain in the beam a distance of 160 meters from the point of origin?

---

2. The Shuttle Takes pictures

The speed of a space shuttle in earth orbit is about 8000m/s. An orbiting shuttle carrying an earth surveying camera photographs North America. The distance from New York to Los Angeles, as measured on earth, is, let us say, exactly 5000km. How much less is this distance when photographed by the shuttle camera? (Neglect the rotation of the earth, as well as its curvature.)

---

3. A space traveler

A space ship is travelling at $0.99c$ with respect to the solar system.

An astronaut of rest mass 50kg steps outside the space ship for a little advance exploration. His pocket rocket exerts a thrust of 5000N, accelerating the astronaut forward in the direction along which the space ship was already travelling, with respect to the solar system.

Expressing velocities as fractions of c, calculate the velocity of the astronaut after one hour has elapsed on the astronaut's wrist watch...

(A) ... with respect to the space ship (Consider the astronaut and spaceship to be a system in which Newtonian mechanics is a good enough approximation) ... and

(B) ... with respect to the solar system.

---

4. Energy of particles

Find the Work that must be done to accelerate an electron from rest to:

(A) $0.9c$

(B) $0.99c$

(C) $0.999c$

(D) $0.9999c$

(E) Find the speed of a proton whose total energy is twice its rest energy.

(F) Find the energy produced from the annihilation of
a positron (mass = m_e; charge = $+e$) and
an electron (mass = m_e; charge = $-e$).

Bibliography

I. In DEPTH

C. Møller, **The Theory of Relativity** Clarendon Press, Oxford, 1952. A classic, still the best no-nonsense comprehensive and rigorous book about relativity. Gathered from notes for a lecture course, the book is oriented to a student of relativity, and is, considering its rigor and subject matter, entirely readable. The flow of logic is discernible and consistent. Not recommended for the beginner or the generalist, not interested in a physics career.

Albert Einstein, **The Meaning of Relativity** Princeton University Press, Princeton, NJ, 1922 revised 1954. Authoritative, typically Einstein, concise, requires some background, not for the non-physicist.

II. For the GENERAL READER

Albert Einstein, **Relativity, the Special and General Theory** *the first edition was published in German in 1916, and has been translated and republished many times, for example:* Crown Publishers, Three Rivers Press, New York, 1961, translation by Robert Lawson. A classic, must-read and re-read. The advertisement, "A clear explanation that anyone can understand" has been withdrawn from the cover, and it's just as well. The book is flavored by Einstein's ability to cut through to the core of a matter, his exuberance at the simplicity of nature, and his stock of clear and telling examples. Above all, the ring of authenticity permeates the book. In his attempt to write what is described as "A Popular Exposition", Einstein has made restrained use of mathematics, with mixed results: in some instances mathematics seems to just pop out from descriptive material full blown, with the logical train of thought left somewhat tenuous. It suffers some from the usual formalistic presentations in which most things follow from the Lorentz transformation without an intuitive chain that permits the mind to accept the reasoning.

Albert Einstein, **The Theory of Relativity and Other Essays** Citadel Press New York 1950, reprinted 1996 by Carol Publishing Group, Secaucus, NJ. Sophisticated, readable (occasional references to higher mathematics can be ignored), remarkable for depth of insight. A varied collection, including the philosophical 1936 commentary, "Physics and Reality."

Martin Gardner, **Relativity Simply Explained** reprinted by Dover in 1997; originally called, "Relativity for the Million," MacMillan, New York, 1962. For what it is, one of the better efforts to do relativity without algebra. Totally phenomenological, but good illustrations. Should be called, "Relativity Simply *Described*;" Explained? No.

John Archibald Wheeler, **A Journey Into Gravity and Spacetime** Scientific American Library, distributed by W.H.Freeman. New York, 1990. This beautifully written book, rich in good illustrations, replete with historical and experimental documentation, tries, in a conversational tone, to take the intelligent reader directly to some of the most sophisticated concepts of relativity, including general relativity, without algebra. It comes as close to achieving that as is probably possible. It argues from examples and analogies, which are disarming, but sometimes deceptively so, and on close scrutiny are not convincing. The book's agenda is tackled anecdotally, and seems frequently without a clear logical thread. The use of classical analogies in particular is sometimes misleading, and, for example, such common-language descriptions of mathematical concepts as, "the camera wasn't where the action was," leads to confusion between look-back time and relativity effects. On the whole, a thoroughly enjoyable and provocative read.

Lillian Lieber, **The Einstein Theory of Relativity** Reinhart & Co. New York, 1936. You won't find this one except in the dusty stacks of an occasional University library. It's a good book if you like your science books written in large type and short lines, a sort of free-form poetry. This book is listed here with tongue in cheek, for its timeless preface: "... just enough mathematics to

HELP and NOT to HINDER the lay reader. ... Many 'popular' discussions of Relativity without any math at all have been written, but we doubt whether even the best of these can possibly give to a novice an adequate idea of what it is all about. ... On the other hand, there are many that are accessible to the experts only..." You said it better than I could have, Lillian.

Leo Sartori, **Understanding Relativity** U. California Press, 1996. This book attempts to be a full-blown text, while also trying to be "readable." It uses good diagrams and lengthy commentary, including a great deal of historical material. It is nevertheless heavy throughout, and never quite succeeds in being what it claims, a "Simplified Approach to Einstein's Theories." For reasons that are not clear, this book is built largely around "paradoxes," like the "car in the garage," or "block in the hole." Leaves the impression that relativity is full of peculiar effects that are never quite made right.

Delo Mook and Thomas Vargish, **Inside Relativity** Princeton University Press, Princeton, NJ, 1987. This collaboration between an English professor and a physics professor attempts to provide the general reader with enough diagrams to explain relativity. The reader should be alert to the fact that the book is about the "Appearance of high speed objects," which includes relativity as well as look-back effects, which are not relativity. The authors find that "a number of things happen at once to introduce several kinds of 'distortions.'" There is an often undifferentiated mix of relativistic and non-relativistic effects. A natural confusion on the part of the beginner is re-enforced rather than clarified.

Robert Resnick, **Introduction to Special Relativity** John Wiley & Sons, New York, 1968. One of the better of the traditional text-book type treatises. It is accurate, thorough, and an excellent introduction for someone who will use relativity in professional activity. The book is heavily algebraic, and somewhat difficult to skim for the student interested in the main ideas without the details. A superb reference book, and except in selected parts is very slow going.

Percy W. Bridgman, **A Sophisticate's Primer of Relativity** Wesleyan University Press, Middletown, Connecticut 1962, 2nd Ed 1983, with a new and substantial introduction by Arthur I. Miller. Another great physicist contributes a unique point of view to relativity. Rigorous, but not heavily algebraic, this book talks through many of the basic puzzles often treated in cursory and cavalier fashion by the classic authors. Just to cite one example: the question of how clocks at different locations are synchronized and what such synchronization means, is discussed in loving and patient detail, much to the reader's benefit.

A. P. French, **Special Relativity - The M.I.T. Introductory Physics Series** W.W. Norton & Company, New York 1968. Not for the casual user, this is for the practitioner, though there are sections which can be read for particular insights. Good descriptions of experiments, and for those with excellent abstraction skills, a fine and thorough text.

Edwin F. Taylor and John A. Wheeler, **Spacetime Physics** W.H. Freeman and Co., San Francisco, 1966. A readable, comprehensive book, organized around numerous examples and problems. Much valuable instruction is done by way of resolving intriguing paradoxes.

Arthur Beiser, *Chapter 1. Special Relativity*, pp 1-38 in **Concepts of Modern Physics**, McGraw-Hill New York, 3rd Ed 1981. By far the best one-chapter survey of the subject with some unique and original derivations. For example, it contains an elementary and intuitively satisfying treatment of the relativistic mixing of electric and magnetic forces, something not usually accessible except through sophisticated tensor mathematics. This chapter is more an exquisite tour of exciting landmarks of relativity than a thorough coverage of all the fundamentals.

Tony Hey and Patrick Walters **Einstein's Mirror** Cambridge University Press, Cambridge, UK, 1997. This is a beautiful and comprehensive book about the science, technology, the history of science and technology, the people and the politics of the twentieth century. As such, Einstein plays a role, but is not the key figure. The book is about quantum theory and the bomb and black holes and light and Michael Faraday and amid lavish illustrations and photographs there are some lavish confusions, such as the chapter title, "Little Boy and Fat Man: relativity in action." On the whole very readable, fascinating, broadening, illuminating. A work of art.

III. SCIENTIFIC BIOGRAPHY

Banesh Hoffman, **Albert Einstein, Creator and Rebel** (with Helen Dukas) Viking Press, New York, 1972. A sparkling biography, weaving insights into the momentous episodes of Einstein's scientific journey into the remarkable story of a successful but troubled and difficult life. A collaboration between two authors who knew Einstein intimately. Hoffman is himself a theoretical physicist who knew and collaborated with Einstein, and has an authoritative and intimate knowledge. Dukas was for years Einstein's personal secretary.

Abraham Pais, **'Subtle is the Lord...' The Science and the Life of Albert Einstein** Oxford University Press, Oxford, 1982. Physicist Roger Penrose called this "... the biography that Einstein would have wanted." Not a book that one reads cover-to-cover, this is primarily the story of the ideas that Einstein contributed to virtually every field of physics, in the context of the other major workers in those fields. Full of the relation of Einstein, the human, to his science, recounted by a profoundly knowledgeable and sympathetic biographer (Example: Einstein's lament, "I wonder if the Lord has played a trick on me," upon his discovery of the equivalence of mass and energy.) Interspersed are episodes in Einstein's life relatively apart from his science, with italicized headings; these were written so as to enable a reader not interested in the more technical aspects of the science to read about Einstein's life, with the opportunity to choose to read forward or backward around these episodes for context. Perhaps the most thorough and complete biography.

IV. HUMAN INTEREST BOOKS

Helen Dukas and Banesh Hoffman, ed., **Albert Einstein, the Human Side** Princeton University Press, Princeton, NJ, 1979. A selection of Einstein anecdotes, letters, and other writings, assembled by Dukas, Einstein's personal secretary for many years, and Hoffman, a physicist and collaborator.

Alan Windsor Richards, **Einstein As I Knew Him** Harvest House Press, Princeton, NJ, 1979. Reminiscences and Photographs taken by the author.

APPENDIX I

In this appendix we will sketch out how Faraday's Law and Ampere's Law combine to predict propagation of electromagnetic waves at a speed determined entirely by the (statically determined) values of ϵ_o and μ_o. All symbols are used as conventionally defined (for example, **E** is electric field, **B** is magnetic field, Φ is flux, i is current).

You are familiar with the integral forms of these laws:

$$\oint \mathbf{E}\cdot d\ell = - d\Phi_B/dt$$

and $$\oint \mathbf{B}\cdot d\ell = \mu_o [\, i + \epsilon_o \, d\Phi_E/dt \,]$$

The expression $\epsilon_o d\Phi_E/dt$ may not be familiar. Imagine that a "current loop" is interrupted by a narrow gap, just as it threads through an Ampere's loop. This interruption, we have so far assumed, would prevent current from flowing. This is not quite accurate in non-steady-state problems, since the interruption is in fact a capacitor, and any changing current (if, for example, the current is alternating) will continue because of the ability of the capacitor to accumulate charge on its two "plates." Thus, in a sense the current continues across the gap in the form of a growing electric field flux. It can be shown easily that the quantity $\epsilon_o d\Phi_E/dt$ is the equivalent of such a current; it is therefore given the name, "displacement current."

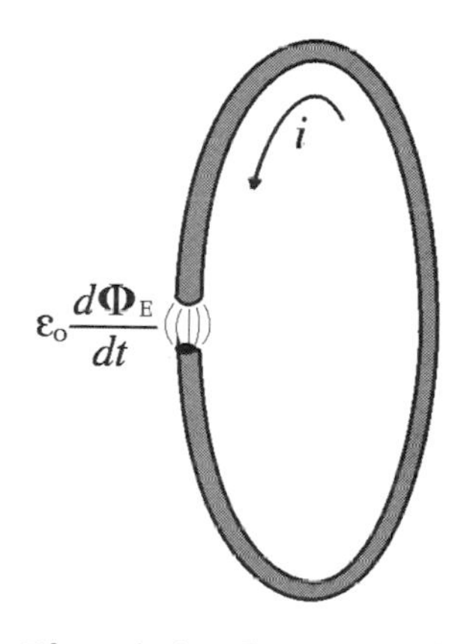

Fig A.1 Current i charges the capacitor at the ends of the wire

It is, in fact, true in general that a changing electric field flux gives rise to a magnetic field in the same way that a current of electric charges does, and thus the form of Ampere's Law shown above is more complete than the form which you first learned.

Physicists tend to be unhappy with descriptions of nature which require adding up a lot of values at different locations, as one does in evaluating a line integral. Thus, $\oint \mathbf{E} \cdot d\ell$ tells only what a lot of different, localized, values of the electric field along a path add up to. Only in examples in which the symmetry is extremely fortuitous, such as when one can draw a path along which **E** is uniform for symmetry reasons, does such an equation tell you what **E** is, anywhere at all.

Calculus attempts to bring average descriptions over finite regions down to exact descriptions in the limit of infinitesimal regions. So, in the spirit of calculus, let us shrink the loops of these line integrals down to infinitesimal size.

Let the loop of a Faraday's Law integral, $\oint \mathbf{E} \cdot d\ell$, be an infinitesimal rectangle, Δx by Δy, in the x-y plane.

Label the four sides of the loop: B (Bottom), T (Top), L (Left), and R (Right). Then designate the average value of the Electric field along any side as $\mathbf{E}_{(side)}$. For example, the average value of **E** along the bottom side will be $\mathbf{E}_B$.

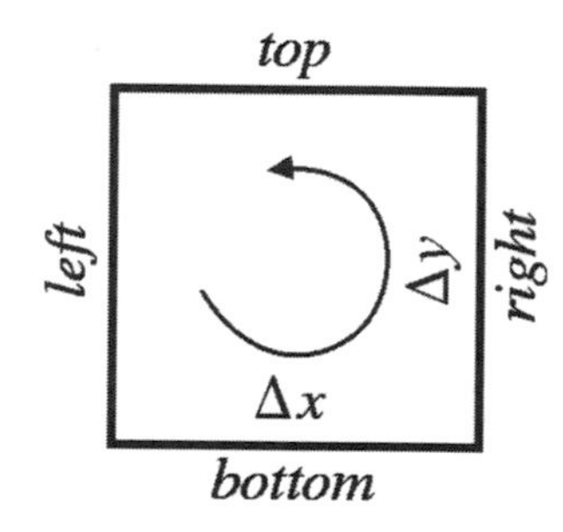

Fig A.2 Infinitesimal Faraday's Law loop, of size Δx by Δy.

The dot product of the integral will select the x-components of **E** along sides B and T, and the y-components along sides L and R. Thus,

$$\oint \mathbf{E} \cdot d\ell = \mathbf{E}_{xB}\,\Delta x + \mathbf{E}_{yR}\,\Delta y + \mathbf{E}_{xT}\,(-\Delta x) + \mathbf{E}_{yL}\,(-\Delta y)$$

$$= (\mathbf{E}_{xB} - \mathbf{E}_{xT})\,\Delta x + (\mathbf{E}_{yR} - \mathbf{E}_{yL})\,\Delta y \qquad [AI.1]$$

$$= -(\mathbf{E}_{xT} - \mathbf{E}_{xB})\,\Delta x + (\mathbf{E}_{yR} - \mathbf{E}_{yL})\,\Delta y$$

If Δx and Δy are small (soon we will go to the limit $\Delta x, \Delta y \rightarrow 0$) then $(\mathbf{E}_{xT} - \mathbf{E}_{xB})$, the difference of the values $\mathbf{E}_x$ between top and bottom can be written as the rate of change of this value in the y-direction $[d\mathbf{E}_x/dy]$ times the y-interval, Δy. And likewise in the x-direction. Thus,

$$\oint \mathbf{E} \cdot d\ell = -[(d\mathbf{E}_x/dy)\,\Delta y]\,\Delta x + [(d\mathbf{E}_y/dx)\,\Delta x]\,\Delta y$$

$$= [(d\mathbf{E}_y/dx) - (d\mathbf{E}_x/dy)]\,\Delta x\,\Delta y \qquad [AI.2]$$

Faraday's law states that this integral, $\oint \mathbf{E} \cdot d\ell$ is equal to the negative of the time rate of change of the magnetic flux through this small loop, $- d\Phi_B/dt$.

This magnetic flux, Φ_B, through the little loop is equal to the surface integral $\oint \mathbf{B} \cdot d\mathbf{S}$ over the little loop. The perpendicular to the plane of the loop is in the z direction, and thus $d\mathbf{S}$ is a vector in the z direction of magnitude $\Delta x \Delta y$. The dot product $\mathbf{B} \cdot d\mathbf{S}$ then picks out the z-component of **B**. Since the loop is small, the magnetic flux can be written in terms of the average value of magnetic field **B**,

$$\Phi_B = \mathbf{B}_z \, \Delta x \, \Delta y \qquad [AI.3]$$

and the right hand side of Faraday's law becomes,

$$- d\Phi_B/dt = - d[\mathbf{B}_z \, \Delta x \, \Delta y]/dt \qquad [AI.4]$$

and, since the loop is fixed, not changing with respect to time,

$$- d\Phi_B/dt = - (d\mathbf{B}_z/dt) \, \Delta x \, \Delta y \qquad [AI.5]$$

Setting this equal to the expression we obtained above for the left side of the Faraday's law equation, [AI.2] for this tiny loop, we get,

$$[(d\mathbf{E}_y/dx) - (d\mathbf{E}_x/dy)] \, \Delta x \, \Delta y = - (d\mathbf{B}_z/dt) \, \Delta x \, \Delta y \qquad [AI.6]$$

from which one can cancel $\Delta x \, \Delta y$ on both sides,

$$d\mathbf{E}_y/dx - d\mathbf{E}_x/dy = -d\mathbf{B}_z/dt \qquad [AI.7]$$

We have succeeded in reducing Faraday's law from Integral to Differential form, a great leap forward, since it deals with values of the fields and their derivatives only at one location.

For loops drawn in the x-z and the y-z planes, one obtains,

$$\begin{aligned} d\mathbf{E}_z/dy - d\mathbf{E}_y/dz &= -d\mathbf{B}_x/dt \\ -[d\mathbf{E}_z/dx - d\mathbf{E}_x/dz] &= -d\mathbf{B}_y/dt \end{aligned} \qquad [AI.8]$$

Using unit vectors, $\hat{\imath}$, $\hat{\jmath}$, $\hat{k}$, one can express the three relations as a single vector equation,

$$\left\{\frac{dE_z}{dy} - \frac{dE_y}{dz}\right\}\hat{\imath} - \left\{\frac{dE_z}{dx} - \frac{dE_x}{dz}\right\}\hat{\jmath} + \left\{\frac{dE_y}{dx} - \frac{dE_x}{dy}\right\}\hat{k} = \frac{-d\mathbf{B}}{dt} \quad \text{[AI.9]}$$

The expression $[dE_x/dy - dE_y/dx]$, for example, represents a local "swirling", or "whirlpooling", a degree of motion round and round the infinitesimal loop in the x-y plane. For this reason the total vectorial quantity on the left of the Differential Faraday's Law expression above, is called the "Curl" of the vector **E**.

You may have noticed that the subscripts remind you of the combinations you encounter in writing vector cross products. The "Curl" of the vector **E** is like the vector cross product of an "operational vector" [d/dx, d/dy, d/dz] with the vector $[\mathbf{E}_x, \mathbf{E}_y, \mathbf{E}_z]$.

An "operational vector" like the one described is referred to as a "vector operator," and can in many ways be treated as if it were a vector. The "operator" [d/dx, d/dy, d/dz] is given the symbol "∇", an upside down delta, and is pronounced "del" (not delta).

The Curl is then the vector cross product of "del" and a vector, and can, in fact be evaluated using the determinant method with which you are familiar:

$$\text{Curl } \mathbf{E} = \begin{vmatrix} \hat{\imath} & \hat{\jmath} & \hat{k} \\ d/dx & d/dy & d/dz \\ \mathbf{E}_x & \mathbf{E}_y & \mathbf{E}_z \end{vmatrix} \quad \text{[AI.10]}$$

Try it, and see that it works. Curl **E** is also written $\nabla \times \mathbf{E}$.

The differential form of Faraday's Law is then, simply,

$$\nabla \times \mathbf{E} = -\,d\mathbf{B}/dt \quad \text{[AI.11]}$$

We should be able to do the same for Ampere's Law. In empty space, where there are no charge currents (i), but only electric field flux currents (displacement currents) Ampere's Law looks like this

$$\nabla \times \mathbf{B} = \epsilon_o \mu_o \, d\mathbf{E}/dt \qquad [AI.12]$$

Equations [AI.11] and [AI.12] are two differential equations in two variables, **E** and **B**. Can they be solved simultaneously? Of course they can.

Differentiate the Ampere's Law equation with respect to time,

$$d/dt\,[\nabla \times \mathbf{B}] = d/dt\,[\epsilon_o \mu_o \, d\mathbf{E}/dt] \qquad [AI.13]$$

Differentiation with respect to space and with respect to time are commutative in this kind of situation, so we can reverse the order of time differentiation and Curl-differentiation, with this consequence,

$$\nabla \times (d\mathbf{B}/dt) = \epsilon_o \mu_o \, d^2\mathbf{E}/dt^2 \qquad [AI.14]$$

Into this expression we can substitute for $d\mathbf{B}/dt$ from Faraday's Law, and obtain an equation with only the Electric Field, **E**.

$$\nabla \times (\nabla \times \mathbf{E}) = -\,\epsilon_o \mu_o \, d^2\mathbf{E}/dt^2 \qquad [AI.15]$$

Exercise 1. Show that

$$\nabla\times(\nabla\times\mathbf{E}) = \begin{vmatrix} \hat{\imath} & \hat{\jmath} & \hat{k} \\ d/dx & d/dy & d/dz \\ \left\{\dfrac{d\mathbf{E}_z}{dy} - \dfrac{d\mathbf{E}_y}{dz}\right\} & \left\{\dfrac{d\mathbf{E}_x}{dz} - \dfrac{d\mathbf{E}_z}{dx}\right\} & \left\{\dfrac{d\mathbf{E}_y}{dx} - \dfrac{d\mathbf{E}_x}{dy}\right\} \end{vmatrix} \qquad [AI.16]$$

Exercise 2. A particular Electric Field has only an x-component, and varies only in the z-direction (this will be the direction of propagation); it does not vary in the y-direction, meaning that $dE_x/dy = 0$.

Show that

$$\nabla \times (\nabla \times \mathbf{E}) \quad = \quad -\hat{\imath} \quad d^2E_x/dz^2 \qquad [AI.17]$$

and that therefore

$$d^2\mathbf{E}/dz^2 \quad = \quad \epsilon_o \, \mu_o \, d^2\mathbf{E}/dt^2 \qquad [AI.18]$$

Mathematicians know that the solution to the equation [AI.18] is of the form,

$$\mathbf{E} = \mathbf{E}_o \sin(z-ct) \qquad [AI.19]$$

where $\mathbf{E}_o$ and c are constants. Differentiate this solution twice with respect to "z",

$$d^2\mathbf{E}/dz^2 = -\, \mathbf{E}_o \sin(z-ct) \qquad [AI.20]$$

Now express the right side of Eq. [AI.18] in terms of the specific solution, [AI.19],

$$d^2\mathbf{E}/dz^2 = \epsilon_o \, \mu_o \, d^2\mathbf{E}/dt^2 = -\, \epsilon_o \, \mu_o \, c^2 \, \mathbf{E}_o \sin(z-ct) \qquad [AI.21]$$

Comparing this relation with Eq. AI.20, one can conclude that the solution $\mathbf{E} = \mathbf{E}_o \sin(z-ct)$ satisfies the equation you derived in Example 2 for any value of $\mathbf{E}_o$ and that the constant "c" satisfies the equation

$$\epsilon_o \, \mu_o \, c^2 = 1 \qquad \text{or} \qquad c = 1 / \sqrt{\epsilon_o \, \mu_o} \qquad [AI.22]$$

If you picture the function $\mathbf{E}_0 \sin(z-ct)$ at time $t=0$ as a sine wave on the z-axis, then at future times any given point on the sine wave (such as the point which is at $z=0$ when $t=0$) moves along the z-axis in such a way that $(z-ct)$ remains constant (zero for the example suggested). "c" thus becomes the velocity, dz/dt, of the point on the sine wave, and is thus the velocity of the whole sine wave pattern.

The conclusion is that "c", which is equal to $1 / \sqrt{\epsilon_0 \mu_0}$, is the propagation velocity of electromagnetic field patterns, known as electromagnetic waves.

APPENDIX II

Proof that in the Minkowski Diagram, $\angle\alpha = \angle\beta$ (ref Chap 16)

This proof refers to Fig 16.5 of Chapter 16, which is reproduced here for reference:

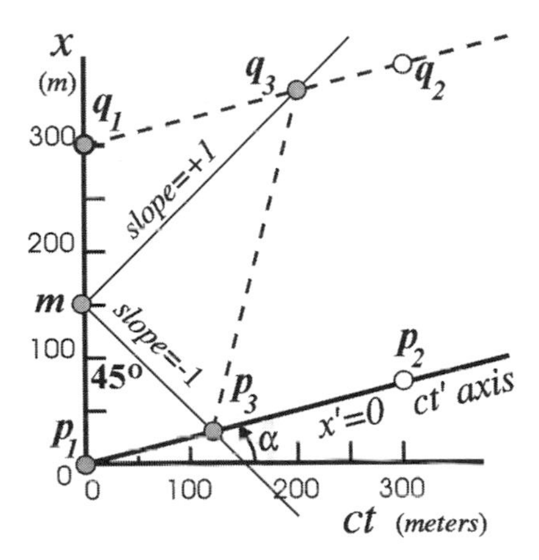

Fig 16.5 Lines from *m* are light pulses; dotted line p_3 to q_3 is a line of constant t'.

In Fig 16.5, p_1 and q_1 are the tail and tip of a rocket. x and ct are fixed axes. The x and ct axes are fixed axes. The rocket has a velocity of ¼c with respect to the fixed axes. The ct' axis is the fixed time axis in the rocket frame. m is the midpoint of the rocket. At $t = 0$, a light flashes at m, sending two light pulses, one toward the tip and one toward the tail of the rocket. The two lines of slope +1 and −1 are the world lines of the light pulses forward and backward. The intersection of the world lines of the light pulses with the world lines of the tail and the tip of the rocket are the events p_3 and q_3. The line p_3q_3 is parallel to the x' axis. It is the burden of this proof to show that the angle that the line p_3q_3 makes with the x axis, called β, is equal to the angle between the ct' axis and the ct axis, called α.

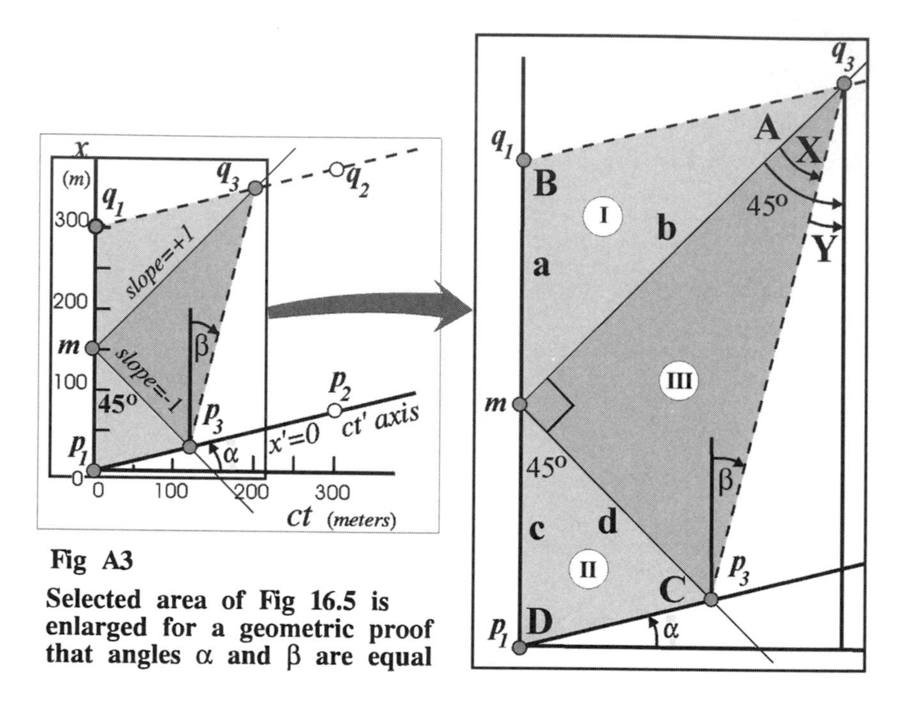

Fig A3

Selected area of Fig 16.5 is enlarged for a geometric proof that angles α and β are equal

I$_3$n ΔI: law of sines: $\mathbf{b}/\sin\mathbf{B} = \mathbf{a}/\sin\mathbf{A}$ *or* $\mathbf{b} = (\mathbf{a}\sin\mathbf{B})/(\sin\mathbf{A})$ [i]

In ΔII: law of sines: $\mathbf{d}/\sin\mathbf{D} = \mathbf{c}/\sin\mathbf{C}$ *or* $\mathbf{d} = (\mathbf{c}\sin\mathbf{D})/(\sin\mathbf{C})$ [ii]

In ΔIII: $\tan\mathbf{X} = \mathbf{d}/\mathbf{b}$;
with [i] and [ii], therefore: $\tan\mathbf{X} = (\mathbf{c}\sin\mathbf{D}\sin\mathbf{A})/(\mathbf{a}\sin\mathbf{B}\sin\mathbf{C})$ [iii]

$\mathrm{A} = 45° - \alpha$ and $\mathrm{B} = 90° + \alpha$ (because $q_1q_3 \parallel p_1p_3$) [iv]
and $\mathrm{C} = 45° + \alpha$; $\mathrm{D} = 90° - \alpha$
therefore: $\sin\mathrm{D} = \sin\mathrm{B}$; $\sin\mathrm{C} = \cos\mathrm{A}$ and so $\sin\mathrm{A}/\sin\mathrm{C} = \tan\mathrm{A}$ [v]

Also: $\mathbf{a} = \mathbf{c}$ (*m* is the midpoint between p_1 and q_1) [vi]

[v] and [vi] in [iii] gives: $\tan\mathbf{X} = \tan\mathbf{A}$ *or* $\mathbf{X} = \mathrm{A}$ [vii]
and because $\mathrm{A} = 45° - \alpha$ [iv] *and* $\mathbf{X} = \mathrm{A}$ [vii], it follows that
$\mathbf{X} = 45° - \alpha$ *or* $\alpha = 45° - \mathbf{X}$

$\beta = \mathbf{Y}$ (alternate interior angles) and $\mathbf{Y} = 45° - \mathbf{X}$; therefore $\beta = \alpha$.

QED

INDEX